How Long is Now?

Fascinating Answers to 191 Mind-Boggling Questions

Questions and answers from the popular
'Last Word' column

Edited by Frank Swain

JOHN MURRAY

First published in Great Britain in 2016 by
John Murray (Publishers)
An Hachette UK Company

1

© New Scientist 2016

A CIP catalogue record for this title is available
from the British Library

ISBN 978-1-47362-859-5
Ebook ISBN 978-1-47362-860-1

Typeset in CelesteST by Palimpsest Book Production Limited,
Falkirk, Stirlingshire

Printed and bound by CPI Group (UK) Ltd, Croydon CRO 4YY

John Murray policy is to use papers that are natural,
renewable and recyclable products and made from wood
grown in sustainable forests. The logging and manufacturing
processes are expected to conform to the environmental
regulations of the country of origin.

John Murray (Publishers)
Carmelite House
50 Victoria Embankment
London EC4Y 0DZ

www.johnmurray.co.uk

Contents

Introduction   1

1. Animal Kingdom   3

2. Above and Beyond   37

3. Domestic Mysteries   52

4. Human Body   73

5. Life on Earth   111

6. The World at Large   136

7. Physics   154

8. Planes, Trains and Automobiles   182

9. Technology   207

10. What the Heck is This?   232

11. A Last Word From Us   241

Acknowledgements   289

Index   290

Introduction

'The Last Word' is all about following your curiosity, wherever it leads. For over a decade, this column has made its home on the final page of *New Scientist* magazine, dedicated to an endless quest for answers. Every week, the *New Scientist* inbox overflows with questions from readers on things that have left them puzzled, and every week we publish the best answers, as supplied by other readers. The fruits of this perennial vine of inquiry have been picked, polished and packed together into the book you now hold in your hands.

Most of the time, the mysteries confounding our readers are drawn from everyday experience: why do wind turbines have three blades? Why does a towel dried on a washing line become stiff? And what causes a hungry stomach to gurgle? Other times, these enquiring minds are focused on more exotic horizons: why do zebras have stripes? Can the sun be extinguished? And what would happen if one black hole swallowed another? Just occasionally, we receive a question that borders on the philosophical: does the sky ever end? And does the sea smell of fish, or do fish smell of the sea?

Our readers are not alone in their search for answers. In this book, you will also find questions that *New Scientist* staff have grappled with in the course of their work, on topics both profound and mundane. If you've ever wondered why ice is slippery, why pollen turns some people into sniffling wrecks but not others, or how life started on Earth, so have we, and our best explanations can be found inside this book. In fact, this quest to entertain life's most thought-provoking questions is a cornerstone of *New Scientist* magazine itself.

No doubt there will be some questions in here that have crossed your minds at one time or another. Others may take you entirely by surprise. But big or small, there's an answer to every question, and sometimes there are even a few. Stay curious.

Frank Swain
July 2016

1 Animal Kingdom

Striped sweater

? Years ago I was told that the black hairs on a zebra heat up while the white hairs stay cooler. This sets up a temperature difference between the stripes, which creates an air flow by convection and helps to keep the zebra cool. Does anyone out there know any more?

Rachael O'Brien
Tamworth South, New South Wales, Australia

The stripes probably serve more than one function, and according to the front-running hypothesis, they serve mainly to repel insects.

Susanne Åkesson and her colleagues at Lund University in Sweden suggest that horseflies are attracted to linearly polarised light. Uniformly coloured animals reflect linearly polarised light, so that makes them a target. Zebra stripes disrupt the polarisation of reflected light, making them more difficult for insects to home in on. Female horseflies need to suck blood in order for their eggs to develop, and biting insects can transmit fatal diseases, so being able to evade them is an advantage.

The researchers tested their hypothesis by standing variously striped and coloured models of zebras, horses and asses in a field. They covered the models in glue, then counted the number of insects that became stuck. The zebras attracted the fewest flies.

Other researchers have lent support to this idea. One theory is that the prevalence of patterning on the hides of such animals increases in proportion to the population of biting insects. It has also been suggested that animals that evolve in areas where tsetse flies are present become striped.

Another suggestion – aside from the stripes' supposed cooling effect – is that they act like a unique barcode, so individuals can recognise each other.

No less a biologist than Alfred Russel Wallace suggested that the pattern provides camouflage. Zebras are most vulnerable when drinking at watering holes, but if they do this at twilight, the stripes merge to a less conspicuous grey. Another take on camouflage is that lions are partly colour-blind, so do not perceive the contrast between zebras and the savannah in the way we do. When zebras are running as part of a herd, the stripes make it harder for lions to pick out individuals.

Mike Follows
Sutton Coldfield, West Midlands, UK

Optic aquatic

? Humans cannot see clearly under water without goggles. How do aquatic mammals solve this problem?

Emma Jackson
London, UK

For light reflecting off an object to be perceived as anything more than dim diffuse illumination, it must be focused on a single point on the light-sensitive retina at the back of the eye. The divergent light rays that strike the front of the eye must therefore be bent (refracted) to varying degrees in order to form an image.

Light is refracted when its waves cross at a glancing angle from one medium to another with a different refractive index. In terrestrial vertebrates light is refracted mainly by the curved surface of the cornea whose refractive index is considerably higher than that of air. The eye's lens has a similar refractive index to that of the surrounding parts of the eye and is responsible only for around one-third of the refractive power of the human eye, serving mainly to adjust the fine focus of the image seen.

Underwater the cornea becomes ineffective as its refractive index is very close to that of water. The underwater world becomes very blurry because light is focused a long way behind the retina and we become in effect very long-sighted. This can be rectified by putting air back in front of the cornea with a face mask or a pair of swimming goggles.

The same obviously cannot be true for animals that live underwater because otherwise their eyes would be of little use. Animals such as fish, cephalopods and aquatic mammals overcome the loss of a refractive cornea underwater by possessing more powerful spherical lenses that can deal with this problem, unlike the lens in the human eye. Next time you eat a fish take out the lens and you will see it is shaped like a marble. The real question is how some animals such as diving birds see clearly in both air and water.

Ron Douglas
Saffron Walden, Essex, UK

Booze cruisers

? I was sitting outside the other evening with friends. Some of us had vodka drinks, others bourbon whisky drinks. There were tiny, slow-moving flies hanging around which we noticed were always attracted to the bourbon glasses rather than the vodka glasses. Does anyone know why?

Zak Friedmann
Austin, Texas, US

The flies are probably common fruit flies (*Drosophila melanogaster*). They are attracted to ripening and decaying fruit, as well as other rotting plant material. The flies are drawn by the complex aroma of esters, alcohols and carboxylic acids that ripe fruits release. Esters are a family of organic chemicals that have distinctive, sweet odours responsible for the characteristic smell of many fruits.

During the bourbon maturation process in the barrel, esters develop, particularly ethyl esters, which give bourbon the floral and fruity aspects of its classic nose. Fruit flies are attracted to this feature of the bourbon bouquet.

Vodka has its fans, but it lacks bourbon's olfactory appeal, as far as fruit flies and whisky aficionados are concerned.

David Muir
Science Department, Portobello High School, Edinburgh, UK

The tiny slow flies are undoubtedly common fruit flies. One summer I had a good colony of them in my kitchen. I first tried to battle them by squashing them one-by-one on the window. Then one morning I noticed a couple of them in a glass containing leftover red wine. I set up a bottle with a few millimetres of red wine inside and it proved successful in capturing them all. Shaking the bottle finished them off.

It's not the alcohol, but the bouquet which attracts them. It is similar to the odour of ripe fruit. That's why they prefer the more aromatic bourbon.

Joachim Koeppen
University of Kiel, Germany

Fly away home?

? Why are some small creatures unafraid of humans? If a bee, ladybird or praying mantis settles on my hand, they seem reluctant to leave. But I have yet to experience lingering contact with a butterfly, small reptile or bird.

Janet Le Page
Johannesburg, South Africa

Far from being afraid of us, there are plenty of creatures – mainly insects – that take the opportunity to drink our blood. Mosquitoes, fleas, ticks and leeches all do this, and we are often unaware that it is happening. Some unfortunate people even play host to the parasitic larvae of the human botfly. Sometimes butterflies land on us in order to drink our sweat for the sodium it contains.

A creature normally avoids getting closer than its flight distance from a potential predator – the shortest head start it needs in order to make good its escape. This means that quicker creatures can often get closer to us. When a housefly lands on someone it may start preening itself, giving the impression of being relaxed. When we try to swat it, the quick-witted fly sees a hand approaching in slow motion and easily evades us.

The same logic applies with larger animals. For example, the flight distance for seals differs depending on whether they are underwater or lounging on a beach. In the ocean, they happily tug at the fins of scuba divers, knowing that they can outswim us with consummate ease. On land, they head for the surf at the first hint of danger.

There are many examples of animals on the Galapagos Islands and elsewhere that allow people to get close because we are unfamiliar and not perceived as a threat. Sadly it was this type of behaviour that led to the demise of the dodo.

Mike Follows
Sutton Coldfield, West Midlands, UK

I feel it is a mistake to anthropomorphise insects. It is unlikely that they have any concept of fear. Each behaves according to how it has evolved and in keeping with its particular attributes. To them, a hand holds no interest, unless they are attracted to its salty secretions, or to any aromatic residues from whatever it has been in contact with.

I don't make a habit of handling bees, but those I have met didn't seem particularly clingy. Yet I would suggest they have had absolutely no past experience in avoiding hands. I once caught one by accident as I walked past a lavender bush swinging my arms. It didn't sting me, presumably because it didn't perceive a threat. Ladybirds cannot readily take off, and they rely on warning colours, together with their nasty taste and smell, to defend themselves from predators. They are therefore not in any particular hurry to fly away.

Flying also takes far more energy than walking. I have had starving ladybirds nibble at my hand, but usually they stroll around, possibly looking for food. When they find none, they crawl up to the highest point and fly away. Mantids are even more unwilling than ladybirds to take flight. They rely on their peculiar way of moving and indeed, not moving, to avoid predation. One taken in hand will clearly show itself to be uncomfortable, and it will try to extricate itself slowly. Different insects respond differently.

Terence Hollingworth
Blagnac, France

Creating a buzz

? If humans stopped taking honey from all beehives, how would this affect the world of bees?

Larry Curle
Huntingdon, UK

As a beekeeper, I would say bees would be little affected. Bees are hardwired to forage for nectar, and if the nectar flow is good they will continue to fill every available space they can with stores of honey. A large store of honey could be useful for the colony to survive a bad year or two.

Wild bee colonies with plenty of available space, such as in a large tree cavity or in a loft space, are capable of accumulating vast amounts of honey – enough to bring ceilings down.

Peter Gandolfi
Chelsfield, Kent, UK

Bees store honey for two reasons: first, to provide food to sustain them during flowerless periods, such as northern winters or long dry periods in tropical areas; and second, to produce a swarm – the only way colonies can reproduce.

If the workers decide it is a good time to swarm, they make a queen cell and have the current queen lay an egg in it. The grub is fed a special diet so that it develops into a queen capable of laying eggs. The old queen then leaves with half of the workers and drones. Before leaving, they ingest about half the stored honey. Beekeepers know just how much honey a swarm can remove – it is amazing that they can still take off.

To prevent their bees swarming, beekeepers have to open each hive every week and destroy any queen cells that have been built. If they fail to do this, then the hive will swarm.

If humans stopped collecting honey, the colonies kept by commercial beekeepers would eventually disappear, and only wild colonies would remain. This would reduce the yield of many crops that use bees for pollination.

Andrew Carruthers
Beaconsfield, Quebec, Canada

Insect impostor?

A brilliant toddler has told me that a butterfly can't be an insect because it has more than six legs when it's a caterpillar. He's wrong but I'm not sure why. What should I tell him?

Lynn Tylczak
Albany, Oregon, US

If your young friend had turned a caterpillar upside down, he would have seen that there are in fact six fully jointed legs at the front end, just behind the head. These will become the legs of the butterfly or moth. The remaining 'legs' have no joints and are really just fleshy outgrowths from the skin called prolegs. Caterpillars have up to four pairs of these plus an extra pair known as claspers at the tail end.

The caterpillars of geometrid moths have no prolegs, just the true legs and claspers. To move along they have to bend themselves into a loop to bring the claspers just behind their legs, and then throw themselves forward as if measuring their own length on the ground. That explains their family name – they 'measure the Earth' with this inching gait – and their popular names: 'loopers' in the UK and 'inchworms' in the US. Watching a looper walking illustrates very well how useful a caterpillar's prolegs are.

Hazel Russman
Harrow, Middlesex, UK

Smelly goat bluff?

? In the forest close to my house in Tuscany there is a group of wild goats. I can smell their pungent odour before I see them. Why are the goats so smelly? And if I can smell them from so far away with my relatively weak primate nose, can a wolf, lynx or another predator smell them from kilometres away?

Alessandro Saragosa
Terranuova, Italy

It's male goats that are smelly, announcing to females that they are virile and have wonderful genes to pass on to offspring. The malodorous chemical cocktail originates from their urine and from scent glands near their horns. It is so potent that it can bring females into oestrous or sexual receptiveness. Japanese scientists put hats on male goats in order to trap and then analyse the stinking volatiles and managed to isolate the most active constituent, 4-ethyloctanal. When female goats smelled this pheromone it triggered ovulation.

For males the reproductive imperative outweighs the risk of predators smelling them. And smelly goats do have ways of avoiding being eaten. They prefer to feed as a herd near steep rocks. Lots of eyes are on the lookout and goats' eyesight is excellent, augmented by having horizontal slits for pupils, an adaptation that improves peripheral vision. If one goat spots danger the whole herd quickly climbs to craggy safer slopes.

Archaeological and DNA evidence agree that the domestic goat (*Capra hircus*) was bred from the bezoar ibex (*Capra aegagrus*) 10,000 years ago by Neolithic farmers in the Middle East. If farmed goats escape they quickly become feral, reverting to the behaviour and instincts of their wild ancestors.

David Muir
Science Department, Portobello High School, Edinburgh, UK

Seal meal

? We hear that polar bears are under threat from climate change and may starve as the world's ice melts. This is because it is difficult for them to venture onto thinning sea ice to find their preferred food – seals. But presumably the seals will still breed somewhere. Where is that, and will it be accessible to polar bears?

Clive Wilkinson
Townsville, Queensland, Australia

That the seals will still breed somewhere is more of a presumption than you might think. Polar bears feed mainly on seal species that prefer thick sea ice, and of these the ringed seal is probably the most important. This species is also the most specialised for breeding in sea-ice caves. I am not aware of ringed seals breeding elsewhere, so if sea ice becomes scarce, that could be a serious problem for them – and accordingly for the bears.

The bears are not limited to a monodiet of seals; they could find other food, ranging from crabs to walrus. However, they have become reliant on a particular annual feeding pattern. They depend on finding a glut of fat-rich marine mammals on sea ice during springtime, then survive the seasonal famine by relying opportunistically on less suitable food such as fish or reindeer, or on scavenging and scraps. Such titbits improve their chances of survival, but would be an insufficient diet on its own. A bear that locates a seal that is not dependent on ice, or a sea lion rookery on a beach, can no doubt make do, but these finds are rare and would not provide sufficient food for many bears.

Jon Richfield
Somerset West, South Africa

The biggest problem polar bears face with their prey is catching it. A bear would get nowhere chasing a seal in the water, and while the bear is faster on land, seals stay close enough to the water for this to be of little use.

Polar bears have therefore adopted a particular hunting strategy. Seals swim under ice to catch fish, but must come up for air from time to time. The bears lie motionless beside holes in the ice and make a quick grab when a seal surfaces. Without ice strong enough to hold the bears' weight, seals could be numerous but the bears wouldn't be able to catch them.

Guy Cox
St Albans, New South Wales, Australia

Wasp-proof

 How do badgers withstand being stung when they raid wasp nests?

A. Gardener
Haywards Heath, West Sussex, UK

Badgers have strong sharp claws, making them powerful diggers. They are in and out quickly. They are also nocturnal. Wasps are reluctant to leave their nests at night and I have heard of people being able to dig them out safely because of this.

Apparently one can remove the whole nest with the wasps stuck inside. This is certainly true of Asian hornet nests, which can be approached closely at night without the hornets reacting in its defence. This is not true for European hornets, I must emphasise.

Terence Hollingworth
Blagnac, France

I observed a badger excavating a wasp nest near Wrexham, UK, on a school biology trip in 1955. The raid was blindingly fast – the badger removed soil from the roof of the nest cavity with its feet, gobbled up much of the brood comb and departed.

The hundreds of disturbed wasps attacked the predator but it had apparent impunity due to its heavy coat, although I suspect the badger's exposed nose received a number of stings. The nutritional value of a comb would be so great that suffering wasp stings would be well worth the pain.

Philip Spradbery
Canberra, ACT, Australia

Preying alone

? Many mammals such as wolves and lions are pack hunters. Many fish and cetaceans also collaborate to herd shoals into bait balls, then eat them. So why are there no flock-hunting birds of prey?

Adrian Bowyer
Foxham, Wiltshire, UK

Cooperative hunting in packs is actually rare among mammals, and seems to occur in intelligent and socially complex species in which the bond between related animals persists into adulthood. Hence it is far more common for mothers and sub-adult offspring to hunt together. The most familiar pack hunters are wolves, lions and orcas. For this type of hunting to be a benefit, the target must offer a more substantial reward than anything within the grasp of a single animal.

Birds are not as physically robust as terrestrial animals, and will not usually target prey larger than themselves, so the cost of sharing normally exceeds the benefit of hunting as a group. It is possible to imagine scenarios in which co-operation would pay off for some bird species, but evolution has no imagination, and cannot plan ahead.

Christine Warman
Hinderwell, North Yorkshire, UK

The American white pelican (*Pelecanus erythrorhynchos*) will gather in flocks to herd fish if the water is shallow. When the water is deeper the birds will usually hunt as individuals.

David Rubin
Ballston Lake, New York, US

My observations on our farm over many years indicate that birds of prey do collaborate and enjoy potential 'meals'. I have seen wedge-tailed eagles work in pairs to stalk wild ducklings in open paddocks. Even though the ducklings were in the care of their parents, and the adult ducks tried to protect their young, they were no match for the eagles.

We regularly observe crows acting in packs to secure their next meal, especially at lambing time. A number of crows will stand a short distance from a ewe with a very young lamb, patiently waiting for either the lamb to wander off from its protective mother or the ewe to wander off without the lamb, for water or to feed. If either happens, they act quickly and attack the lamb. I have also observed dozens of crows in a flock chasing small mobs of ewes and young lambs in open paddocks, trying to separate them and attacking any lambs that lag behind. Is this why a group is called 'a murder of crows'?

Anna Butcher
Brookton, Western Australia

I visited a falconry near Perth, UK, a few years ago and was introduced to Oscar, a Harris's hawk that was adept at working with novice falconers. A falconer told me that these hawks are unusual in that they hunt in flocks, and that this might be the source of their flexibility and social intelligence in working with humans. I recall walking up a hill and being surprised to find Oscar walking beside me. I was told he probably only wanted a lift to the top, and I let him stand on my glove. When I got to the crest he took off again. He was obviously as tired as I was.

Robert Williams
East Croydon, Surrey, UK

Your questioner is mistaken in believing that there are no flock-hunting birds of prey. The Harris's hawk (*Parabuteo unicinctus*) is well known to hunt in family groups of up to fourteen, but typically only five or six, in much the same way as the land mammals cited. This practice suits birds hunting in the desert and dry scrub landscapes of the hawk's typical habitat.

Harris's hawks ambush prey by chasing their quarry into a group of waiting birds. During the chase, any hawk can take over when the leader gets tired or is thrown off course. They are reputed to be the most successful predators on the planet in terms of ratio of kills to hunting attempts, and share their spoils among the group. Because of a gregarious nature, they have been widely adopted as falconry birds – they are relatively easy to breed in captivity and can be trained to attack a variety of game. They are sometimes flown by falconers in a 'cast' of two birds, or even three, to take advantage of their natural behaviour.

David Ridpath
Drybrook, Gloucestershire, UK

Merlin's magic

? My dog, a four-year-old whippet called Merlin, occasionally seems to select and eat goosegrass (*Galium aparine*) from a hedgerow, carefully pulling out stalks and ignoring the other plants surrounding it. Is he self-medicating, and if so, how does he know which plant to eat?

Andy Maslen
Salisbury, Wiltshire, UK

The apparently self-medicating behaviour of animals is known as zoopharmacognosy, and your dog is probably eating the goosegrass to induce vomiting in response to a mild stomach irritation. My cat does the same thing, often selecting specific species of grass. She did this both in the UK and now in Hong Kong, so the species seems irrelevant beyond individual preference. Goosegrass is so-called because geese like to eat it, purportedly because it helps to eradicate worms from their digestive tract – which folk tales explain by suggesting the plant's tiny hooks catch worms and carry them from their bodies.

Peter Sims
Hong Kong

As the owner of a four-year-old whippet that shows no interest in goosegrass, I was curious to find out what may be stimulating Merlin's predilection. Goosegrass, also known as cleavers, contains many chemicals that may have a biological effect, including asperuloside, caffeine, and phenolic compounds. It has been used as a tea by herbalists and is said to act as a diuretic and 'lymphatic'.

Its appeal to dogs is noted on the online Gundog Training Forum, where a number of contributors report that their dogs actively seek it out. The effects of goosegrass are obviously open to question. There is an unverified report that it lowers blood pressure in dogs without affecting their hearts or producing unwanted effects. But I doubt any dog would either be aware of its own blood pressure or actively seek to lower it. Because goosegrass belongs to the same family as coffee, could it be that it gives dogs the equivalent of their owner's mid-morning caffeine boost?

As for how your correspondent's dog finds the goosegrass,

we all know that dogs can distinguish smells way beyond our imagining. Phenols have a very strong scent that even humans can detect, so it is probably very easy for dogs to find, even if the phenols are present in small quantities. Still, as the Gundog Training Forum points out, there is no substitute for visiting a vet to check whether this behaviour should be encouraged. Just because something is eaten without apparent harm does not mean it is good for you. As one vet said to me about a substance now widely recognised as being harmful to dogs: 'Have you ever seen a dog refuse chocolate?'

Gillian Coates
Anglesey, UK

Rose nose

? My golden retriever is 'Dudley-nosed', meaning it has a pink nose rather than a black one. Such noses were first noticed in bulldogs from Dudley in Worcestershire in the UK, hence the name. If this is a genetic mutation, how could it jump to other pure-bred dogs?

Alan Moskwa
Joslin, South Australia

Dudley nose is caused by a mutation in the *TYRP1* gene that affects the synthesis of the pigment that causes dogs' noses to be black. Although first noted in bulldogs, it appears in other breeds as well.

Dudley nose only occurs when a dog is homozygous, or has two recessive alleles for the mutation. This means a pair of dogs could have black noses but be carriers of the mutation and so have Dudley-nosed puppies.

Dog breeders are encouraged not to breed Dudley-nosed dogs, but there is no way to stop carriers being used in breeding, unless DNA testing is carried out. This is how Dudley nose can spread to puppies among pure-bred dogs.

Nicky Rudling
Waltham Chase, Hampshire, UK

Cats' eyes in sight

[?] The pupils of domestic cats' eyes are ovoid but those of big cats such as lions and leopards are round like human pupils. Why is this and do the differences confer any advantage either way?

John Neimer
Weymouth, Dorset, UK

For efficient light-gathering, the eyes of nocturnal vertebrates such as small cats have large pupils relative to the focal length of the eye. This results in a narrow depth of field of clear focus. What's more, biological lenses focus different colours by different amounts, causing chromatic aberration (where different colours are not focused equally) and blurring. The coloured fringes in images produced by poor-quality optical equipment are caused by this same problem.

To overcome these difficulties, evolution has equipped some animals with multifocal lenses formed of concentric rings with different focal lengths. Each ring focuses a different part of the spectrum, giving a sharper image. A slit pupil is a later adaptation, so that when an ovoid iris constricts in brighter light, the full diameter of the lens can still be used. A circularly constricting iris would cover the outer parts of the lens.

Why do animals active during the day like lions and ourselves have monofocal optics and circular pupils? We have a small pupil that gives a large depth of focus and lenses of longer focal length that minimise the effect of chromatic aberration. A reasonable hypothesis would be that multifocal optics in diurnal vertebrates would give little if any benefit, so didn't evolve.

David Muir
Science Department, Portobello High School, Edinburgh, UK

In almost all vertebrates, the pupil is large and round in low light but constricts as light levels rise. Although the constricted pupil is round in most species, in some animals it can be a vertical or horizontal slit, or even, as in some geckos, composed of several tiny pinholes. The cuttlefish has perhaps the most exotic pupil shape: it becomes a W in bright light.

In low light the primary requirement is to be able to see anything at all. This entails the eye gathering as many photons as possible, which requires a large pupil. However, there is a disadvantage to this, namely reduced image quality. The main reason is that when the pupil is at its widest, the entire lens – lying behind the pupil – is being used to form the image.

Light going through the edges of the lens ends up being focused more than light going through the centre, an effect known as spherical aberration. Not all of the rays are focused at the same point, resulting in a blurry image. In low light however, this is a price worth paying. At high light levels the pupil constricts to ensure only a small part of the lens is used, resulting in better image quality.

In humans and other animals with round pupils, a

sphincter muscle at the tip of the iris constricts to form the pupil. The disadvantage of a circular muscle is that the pupil cannot be closed down beyond a certain size. In animals with a slit pupil, the corresponding muscle runs along either side of the slit. Consequently, although the constricted human pupil and the dilated pupil differ in area by a factor of about 16, a cat's pupil changes by a factor of at least 135, and the pupil area can approach zero in bright light. Slit pupils are widespread in nocturnal animals whose retinas are adapted for low light and might be damaged in bright conditions.

Ron Douglas
Saffron Walden, Essex, UK

Dig up the truth

? We noticed a huge number of molehills when walking through fields and woods on the Thames Path in Oxfordshire recently. Most were on or near the path rather than out in the fields. Why is this?

Catherine Kennally
London, UK

I've only ever found one mole out in the open, which is unsurprising because of their lifestyle. In woodland and in dense vegetation they can remain concealed and find food by scrabbling about under the leaf litter. There is no need for them to waste energy digging tunnels. However, on open ground, trying to feed on the surface would leave them dangerously exposed and there is little to eat there anyway. Such areas, especially paths, represent a barrier best negotiated underground. By digging a network of tunnels – the cause of molehills – they can move unthreatened beneath the path and, as they do, hoover up any worms or insects that find their way into what are effectively pitfall traps.

Terence Hollingworth
Blagnac, France

Moles are probably very active underground beside the Thames Path because they are good at avoiding being flooded out, and so favour the higher ground and slopes which the path follows rather than low-lying land. Humanity chooses these areas for paths partly to keep our own feet dry. I recall *New Scientist* reporting that if threatened by flooding, moles tunnel up to the surface, can swim, and can sense their way somehow to adjacent dry land despite being nearly blind.

John Forrester
Edinburgh, UK

Dung addicts

? We muck out our horse paddock twice a day. In the late summer and autumn there are usually one to three dark-blue iridescent beetles sitting in burrows beneath the horse droppings. How do they get there in such a short time, which is perhaps twelve hours at most? Do the beetles come through the soil (which is often very hard and dry in the summer) or do they travel over the ground?

Ilka Flegel
Kapellendorf, Germany

Some years ago I released five species of dung beetle onto our farm in southern New South Wales. These species originated in South Africa, and after years of testing were allowed into Australia along with other species. The reason that the beetles arrive on the dung so quickly is that they are strong flyers, some with quite a loud buzz. Eggs are laid in cowpats and the larvae hatch in about three days, and dig tunnels pulling the dung down. This kills any fly larvae that are present, which take five days to hatch. A fly can lay about 3,000 eggs on one cowpat, so the improvement of my farm pasture and the reduction of fly numbers have been striking.

David Hamilton
Sydney, Australia

The beetles burrowing in the horse droppings have almost certainly flown there, attracted by the smell. They are likely to be dung beetles of the family *Scarabaeidae* (scarab beetles). Dung beetles are loosely divided into rollers, burrowers and dwellers. Rollers will roll a ball of dung away and bury it for food or egg-laying. Burrowers dig down into the dung, feeding and breeding inside it. Dwellers live on top of the dung.

Your beetles seem to be burrowers. Most dung beetles have an exquisitely sensitive sense of smell and are powerful flyers. To avoid competition, and lessen the risk of the dung drying out and becoming impossible to mine, the beetles arrive very quickly. I worked for twenty years in Zambia and frequently had to take trips into the bush. On one occasion I was driving alone on a remote road and pulled over to relieve myself. As I crouched there was a loud buzzing and when I stood up I saw a dung beetle had arrived, and was already hard at work expertly forming a portion into a ball. Having done so the beetle hurried away with it. As I watched grimly fascinated, another arrived on the wing and then another and another.

I was so astonished that I didn't think to count how many there were, but within thirty minutes there wasn't a trace of my faeces left. Still more dung beetles flew in, presumably attracted by the residual smell, and wandered around forlornly. It was one of my more memorable wildlife experiences.

Alistair Scott
Gland, Switzerland

Off colour

? I spotted a blackbird in the garden, which is not black but light grey, and it did not have pink eyes so I guess it is not an albino. It spread its wings and lay in the sun; in due course it flew off. I've never seen a blackbird with this colouring before. Can anyone tell me more about it?

Eric Bignell
Southwell, Nottinghamshire, UK

I am curator of the Bird Group at the British Natural History Museum, and heritable colour aberration in birds has been my main research area for the past fifteen years. During that time, I have learned that the identification and naming of colour aberrations still present a problem in the ornithological world.

The main reason for this is probably that the appearance of similar aberrations, which are caused by genetic mutations, can differ radically between birds of different species, depending on their original pigmentation. It also varies with sex and age. This plus the fact that a bird might be too far away or moving too quickly to see clearly often makes it hard to distinguish between different aberrations.

A variety of names are used seemingly at random to identify

the mutations. The terms albino and partial albino are the most commonly applied, and are most frequently wrong. Your questioner correctly assumed the bird is not an albino because it does not have pink eyes. However, even without seeing the eyes you can tell this bird is not an albino because it is grey and therefore has some melanin pigment.

Animals with the genetic mutation that causes albinism cannot produce any melanin because the enzyme tyrosinase is absent in the pigment cells. So there is in fact no such thing as a partial albino. This blackbird's grey plumage is the result of an aberration called dilution. From the Latin *dilutior*, meaning paler or weaker, dilution can be defined as a reduction of melanins. The number of pigment granules is reduced but the pigment itself is not changed. The lower concentration of granules forms a weaker or diluted colour. This is analogous to a black-and-white photograph in a newspaper: a high concentration of black ink dots close together is perceived as black while fewer black dots in a similarly sized area appears grey.

Hein van Grouw
Tring, Hertfordshire, UK

Further to earlier answers, it appears that the off-black blackbird is 'anting'. Some birds deliberately lie with their wings outstretched on top of an ants' nest so that the insects and the formic acid they release kill parasites clinging to the bird's feathers. Anting is quite unusual, and many of your readers may have never seen it, or recognised what is happening. Birds often stay in this position for some time, giving people the opportunity to both observe and photograph them.

Andrew Carruthers
Quebec, Canada

The reason given by an earlier correspondent from Canada for the off-colour blackbird's behaviour seems questionable. In my garden in the south of England birds often show this behaviour. The species involved include blackbirds, robins, sparrows and pigeons, and all behave the same way: wings spread and tail raised with their back to the sun. They choose various parts of the garden and seem only to do this on sunny days.

I've not seen any evidence of the involvement of ants killing parasites on the birds' feathers as your correspondent suggests. This behaviour is frequent in the late spring, often involving more than one bird at a time in different parts of the garden. Is it possible that they are just enjoying the warmth of the sun? The stretching, the closed eyes, and the orientation certainly suggest this. I haven't seen first hand the behaviour of ants in Canada as your correspondent obviously has, but perhaps they are just taking advantage of the temporary shade provided by the blackbird's wings.

Mike Garrett
Woking, Surrey, UK

Further to previous answers, I think this bird is sunbathing, not 'anting'. The ultraviolet B-rays in sunlight facilitate a crucial step in the biosynthesis of vitamin D. In humans, this occurs directly in the skin, but in birds the skin is shaded by feathers. To resolve this, birds use an oil secreted by the uropygial gland or preen gland near the base of the tail. This contains a precursor steroid that is converted into vitamin D by sunlight. The bird spreads this oil on its feathers by preening, then suns itself either as described previously, or more briefly in flight, and consumes the photosynthetic product with the next preening. Perhaps this prolonged sunning behaviour is seen mostly in the spring because of a vitamin D deficiency acquired during the dark winter.

Charles Sawyer
Byron Bay, New South Wales, Australia

Clucking on

? Why do hens cluck loudly after laying an egg?

James Turton
Bristol, UK

I have kept domestic fowl free range for almost sixty years, namely bantams and silver-laced Wyandottes. This is what I have observed: a hen enters the nest ten to twenty minutes before laying and settles herself. She stands to lay the egg which is soft-shelled and hardens on contact with the air. She then sits quietly for a few minutes, preening, crooning and resting. The hen then jumps up from the nest with a loud cackle and given the freedom of the range, runs a considerable distance. The dominant rooster hears the cackling, runs with wings outstretched to the hen, and mates with her immediately. He then performs a 'stamping' display and both go off happy.

So it seems that the cackle helps to attract the rooster to mate once an egg is laid. If there are rival males, the rooster will patrol outside the nest like any expectant father and then mate with the hen as soon as she has finished laying. In that case, the cackle may be quiet to non-existent. When mating at other times, the stamping display comes beforehand and there is no cackling.

Nina Dougall
Malmsbury, Victoria, Australia

Perfect perch

? Birds perch standing up, bats upside down. Are there any bird or bat exceptions to this? And why do the two perch differently?

David Hambling, via email

There is a misconception that bats can't take off from an upright position, owing to their small leg bones and muscles which have been reduced to make flight more efficient. However, flapping their tail membrane allows bats to launch upwards. Granted such an ungainly take-off would make them vulnerable to predation if they nested on the ground, which makes dropping into flight from a perch a better strategy.

It's not just risk of predation that explains the difference though. Bats also have superior aerobatic ability. These animals can invert and come to a virtual standstill in flight which allows them to grasp a suitable roosting position from underneath. This means bats can monopolise the ceilings of caves and other inaccessible roosting sites. They are more manoeuvrable because their wings are larger relative to their body mass than is the case for birds. In addition, though wings have evolved from arms in both cases, the bone is limited to the front edge of a bird's wing, whereas in bats the fingers extend across and to the rear edge of the wing, giving bats finer control of the wing surface.

Conversely, birds' lack of manoeuvrability might limit them to more accessible roosts. And this may help explain why birds don't sleep as we understand the term – they rest each side of the brain in turn so that they stay alert to predators, and possibly to avoid falling off their perches.

However, there are a few birds including the vernal

hanging parrot (*Loriculus vernalis*) that roost upside down in trees. The tendons in bats' and birds' feet are arranged to close the claws and lock the feet to the perch when the creature is relaxed, minimising energy expenditure. If a bat dies in its sleep, it doesn't automatically fall to the ground, and needs to be knocked off its perch.

Mike Follows
Sutton Coldfield, West Midlands, UK

Upright snooze

? Why do some animals, such as humans, lie down to sleep, whereas others, such as elephants and giraffes, stand?

Martie van der Walt
Pretoria, South Africa

There are several factors that affect an animal's sleeping style: how easily it can get up and stay up, how easily it can lie down, how comfortably it can stay down and what kinds of threats it faces in its environment. In general, the larger a land animal, the less it likes getting up or staying down, so very heavy animals such as elephants don't lie down much. It is hard for recumbent elephants to breathe or to rise suddenly. Instead, their legs are adapted to support their weight vertically, so that a healthy elephant can stand indefinitely and sleep lightly, possibly leaning against something. It will only lie down for a deep sleep when it feels secure enough.

Giraffes sleep even less; they are more vulnerable to predators. A cat, pig or goat can take off quickly from a sleeping start, so they are free to lie down and curl up. Cattle and horses are somewhere near the break-even point. They keep to their feet as a rule, but can get tired enough to rest, or nap while lying down. In the wild, however, they rarely lie down when they are nervous of predators.

Jon Richfield
Somerset West, South Africa

As a qualified field guide with extensive experience of elephants in the wild, I can assure your readers that they sleep lying down, contrary to popular belief.

Rod Murphy
Pinegowrie, Gauteng, South Africa

Spider woman

? **My wife is often bitten by spiders. Can anyone explain her allure to arachnids, and suggest how she might reduce it?**

Steve Yorke
White Rock, British Columbia, Canada

My sympathy to the correspondent's wife, but any complaint of repeated spider bites is more or less certainly misdiagnosis. Spiders rarely bite, and only when people handle them roughly.

Unfortunately, people commonly reject suggestions that their symptoms, however genuine, have no connection with spiders. To compound the problem, medical staff with poor arachnid knowledge, or who lack the patience to argue, sometimes indulge the fiction, reinforcing the person's readiness to attribute every localised idiopathic skin condition to spider bites.

The range of infections and injuries that people mistake for spider bites is startling: it includes trivial strep pustules, recurrent MRSA infections, mycobacterial abscesses, viral vesicles, chemical burns, allergic rashes and, of course, other kinds of bite – from mosquitoes, blackflies, bedbugs, fleas, lice, ticks or blood-sucking mites.

My neighbour recently complained of repeated bites from jumping spiders. Once he accurately described the bites, it was clear they were from our local *Ceratopogonidae* (known as no-see-ums or biting midges).

People vary dramatically in their reactions to bites, and *Ceratopogonidae* are likely suspects for the bites incurred by your correspondent's wife. As for the questioner's apparent immunity, it might simply reflect a lack of reaction, if bites are in fact the cause.

Jon Richfield
Somerset West, South Africa

Holocene park

? Dinosaurs died out some 65 million years ago and mammals exploited the vacant niche. If they had not died out, which branch of dinosaur might have evolved to become the dominant, most intelligent creature?

Jack Harrison
Tobermory, Isle of Mull, UK

Birds evolved from theropod dinosaurs during the Jurassic period and species that survived the extinction event 65 million years ago are ancestors of those we see today. Coincidentally, I was watching a flock of geese feeding by a pond recently and it occurred to me that they could be mistaken for a pack of velociraptors.

Were it not for the extinction event, the 'terrible lizards' might well still dominate. In the absence of environmental change, there is little selective pressure to drive the evolution of new species.

The climate change associated with the end of the Cretaceous period would have fired the starting pistol for an evolutionary race, with some species undergoing significant adaptive changes to fill the new niches.

This is akin to shuffling a deck of cards. It is impossible to work out in advance what species might dominate or even what a new species will look like, because it is impossible to know what combination of traits will be more favourable than another. Besides, sometimes it is enough just to be the first to occupy a particular niche.

Mike Follows
Sutton Coldfield, West Midlands, UK

Back in the early 1980s, palaeontologist Dale Russell suggested that the dinosaurs most likely to evolve intelligence would have been the troodons. These were a family of moderately sized two-legged predators thought to have been closely related to birds. At the time, I wrote an article for *Omni* magazine about how troodons might have developed. We have learned a lot about dinosaurs and their evolution since then, and Russell's original idea of an upright and rather human-looking 'dinosauroid' no longer seems realistic.

Palaeontologist Darren Naish gives some alternatives from humanoid-looking species to almost-science-fiction-inspired bioraptors and brings the story more up to date online at scienceblogs.com (*bit.ly/CleverDinos*).

One other thing we have learned about dinosaurs and their evolution is that birds did evolve from dinosaurs, as Thomas Henry Huxley had suspected after he saw *Archaeopteryx*.

The modern system of classifying all animals on the basis of their evolutionary relationships, called cladistics, thus considers birds to be dinosaurs. So, strictly speaking, dinosaurs did not die out but survived as avian dinosaurs, otherwise known today as birds. Crows and ravens have also evolved some degree of intelligence.

Jeff Hecht
Auburndale, Massachusetts, US

Dino DNA

? I recently watched the movie *Jurassic Park* and its sequels, in which DNA from the stomachs of mosquitoes stuck in amber was used to recreate the dinosaurs. Could you do this in real life?

Joseph Scott (aged ten)
Stonehaven, Aberdeenshire, UK

Insects appeared on Earth together with plants around 480 million years ago, but the earliest record of blood-feeding insects is much more recent. A mosquito found encased in rock in Montana is reckoned to have lived 46 million years ago, and another blood-feeding bug discovered in China predates that mosquito by 30 million years. This falls within the early Cretaceous, between 142 and 64 million years ago, which is when *Tyrannosaurus rex* and most other well-known dinosaurs lived, making their coexistence with blood-sucking insects highly likely.

The conclusion that these organisms fed on other animals'

blood came from analyses of their abdominal contents. These identified iron-containing haem – the main component of blood, which carries oxygen and gives red blood cells their colour. The blood meal would have contained mainly red blood cells and some white cells. In mammals, red cells lack a nucleus containing DNA. All other vertebrates have red blood cells with a nucleus, each one containing the whole genome. So the DNA content would be much higher in a blood meal from a dinosaur than in one from a mammal.

However, the high level of degradation meant that no whole blood cells were identifiable in the recovered fossil material. Both digestion of the blood and fossilisation of the insect would almost certainly have destroyed any DNA. In addition, DNA has a half-life of approximately 500 years, and would be too broken to contain any information after about 1.5 million years. So finding a complete, intact genome in a dinosaur fossil or in an insect that fed on it is highly unlikely.

Let us imagine, however, that somehow an insect was embedded in amber just after sucking blood from a dinosaur, and preserved in super-ideal conditions. The owner of the extracted DNA could only be pinpointed by sequencing it and matching it to a reference genome. But we have no reference genome for a dinosaur, so there would be no option but to try it and see what comes out.

For that, we would need a whole, undamaged nucleus to transplant by injecting it into, say, an unfertilised frog egg with its own nucleus removed. The frog egg would provide everything necessary for development, and the transplanted nucleus would provide the genetic instructions for the new organism.

What happens then? A frog is an amphibian and does not make hard-shelled eggs, so would the transplanted dinosaur nucleus develop? Who knows? The Jurassic Park movies may have brought back to life the most amazing animals that ever walked the earth, but I wouldn't recommend trying it at home.

Alena Pance
Wellcome Trust Sanger Institute, Genome Campus, Cambridge, UK

2 Above and Beyond

Clouding the issue

[?] Are there any wavelengths at which the sun still casts a shadow when the sky is full of clouds? Could I make a sundial that would work on a cloudy day?

Stephen Parish
London, UK

One way to work out the position of the sun – and thus to deduce the time – when conditions are overcast is to observe the polarisation of what light is available. This phenomenon, where the light waves oscillate in the same plane, is something that insects and birds exploit for navigation.

In general, scattered light is polarised at right angles to the sun. So when the sun is at its highest point, light is close to being horizontally polarised along the entire horizon. When the sun sets directly west, the sky will be vertically polarised at the horizon due north and south.

In 1848, English inventor Charles Wheatstone presented the 'polar clock', a sundial-like device that could be used when it was cloudy. By angling the tube towards the North Pole and turning a prism in the eyepiece until the light vanished, the relative angle of polarisation of available daylight could be deduced, giving the position of the sun and thus the approximate time. It has also been suggested that Vikings used crystal sunstones to locate the sun when it was obscured by clouds or just over the horizon.

Mike Follows
Sutton Coldfield, West Midlands, UK

Some radio waves can go through clouds, and the sun does emit radio waves, so it would be possible to build a sundial that works on shady days. But you would need a sundial that was large in comparison to the wavelength of the radio waves in question, otherwise the waves would simply refract around it and you wouldn't get a shadow. The shadow would have to be detected by a huge array of antennas designed to pick up the sun's radio waves. This is not simple. It took a long time before radio waves from the sun could be detected, and it was not until 1942 that English physicist James Stanley Hey managed it.

There is another option though. X-rays are also emitted by the sun and these would penetrate clouds too. It might be easier to build a sundial based on X-rays because there wouldn't be the problem of diffraction and you could see where the X-rays fall using a fluoroscope.

Eric Kvaalen
Les Essarts-le-Roi, France

Sky lie

When I went on holiday from the UK to Australia, I noticed that the full moon looked different to how it appears at home. Online, I found many comments about the moon in the southern hemisphere appearing to be 'upside down'.

On my return home I compared photographs that were taken in Northern Ireland with those I took in Cairns in Australia, and there seemed to be only about a 70° difference rather than 180°. So does the face of the moon have any variation locally in terms of the aspect facing Earth, and are there two places in the world that when compared would give a 180° difference?

Graham Finney
UK

This raises several interesting issues. To start with, consider observers standing at latitude 54° north in Northern Ireland, looking up at the full moon at midnight, when it is highest in the southern sky. The moon will be aligned with its south pole towards the southern horizon.

Now suppose you could transport them into the southern hemisphere over the course of a few minutes. As they approach the equator, they would see the moon rise higher and higher in the sky, but with its orientation unchanged.

As they cross into the southern hemisphere, the moon would be at the zenith, vertically overhead. Thereafter they would have to bend over backwards to see the moon in the same orientation – the 'right way up'.

Eventually they would have to stand up to avoid falling over backwards. Turning around to look at the moon now, they would see it in the northern sky, with its north pole towards the northern horizon. In that sense, the moon is now perceived to be upside down, but really it is the observers who are 'upside down'.

Now consider the moon's path in the sky. The moon (more or less) follows the arc of the ecliptic, the apparent path of the sun across the sky, and the face of the moon is (more or less) at a constant orientation to the ecliptic. The angle the ecliptic makes with the horizon depends on the latitude of the observer and the time of year.

During the equinoxes, it turns out that at the equator, the ecliptic rises vertically and passes through the zenith. For our observers in Northern Ireland, the plane of the ecliptic is at 54° to the perpendicular, towards the southern horizon; for the observers in Cairns (latitude 17° south) it is at 17° to the perpendicular, leaning towards the northern horizon.

This means that photos of the moon near to moonrise taken in the two places would show a difference of 71° in the orientation of the moon's face relative to the horizon. I think this is the explanation for the apparent rotation your questioner observes. It is because the photos were taken when the moon was low in the sky, and because Cairns lies roughly 71° south of Northern Ireland in latitude.

I mentioned earlier that the face of the moon is 'more or less' aligned to the ecliptic. This is because the moon does not keep exactly the same face towards the Earth; it wobbles, a process known as 'libration'.

Partly this is due to the moon's orbit not being exactly circular. The moon slows down or speeds up as it goes around the Earth, so its orbital speed does not keep step with its rotation. At some points, it is possible to peer a little way round the moon's eastern or western limb.

In addition, there is a daily libration. This is the result of Earth's rotation carrying the observer from one side of the Earth-moon axis to the other, bringing a tiny sliver more of the moon's surface into view in the process. For the same reason, the observers based in Northern Ireland would see a little way over the moon's north pole, and the observers in Cairns slightly over its south pole.

Finally, the moon's orbit is inclined slightly to the ecliptic, by about 1°, which brings the north or south of the moon a little more into view at various points in its orbit. The combination of these effects makes it possible to see about 60 per cent of the moon's surface at one time or another, rather than exactly half, as many people assume.

This means, of course, that two photographs of the moon taken some time apart in different places would not show exactly the same face, regardless of its orientation.

David Walmsley
Wokingham, Berkshire, UK

Lunar loser

? If the moon were to disappear, how long would it take for the tides to stop?

Martin McCann
Worcester, UK

If the moon suddenly disappeared, the main force that causes the tides would stop immediately (or just after the time it takes for gravity to travel from the moon to Earth – about 1.5 seconds). But because the seas would be piled up on two opposite parts of Earth, and would be lower in the belt between those parts, the water would start to oscillate. Initially it would flow from the two 'hills' of sea towards the low-tide area where it would pile up again. Then the water would stop and start flowing the other way. This oscillation would in theory go on forever, but in reality would become more chaotic through friction and the interference of the land and would diminish in amplitude over time from hydraulic drag on the seabed.

Eric Kvaalen
Les Essarts-le-Roi, France

The good news is that it would take a very long time for the tides to stop – until Earth stops spinning and has only one face pointing at the sun, or the oceans boil away. The reason is that there are two celestial bodies that cause tides on Earth: the moon and the sun. The strength of the sun's gravitational effect on tides is slightly less than half that of the moon. The two bodies can either act together or against each other depending on where the moon is in its orbit.

The phenomenon of the spring tide is produced when the sun and moon line up, giving exceptionally high and low tides. Neap tides – when there is the smallest difference between high and low water – occur when the tide-generating force of the sun opposes that of the moon. The magnitude of tides produced in the seas of a moonless Earth would only be slightly less than the current weakest neap tide.

Peter Skelly
Bedford, UK

Drowning sun

 Could all the water in the universe put out the sun?

Maya (aged six)
San Mateo, California, US

Fire is a chemical reaction that needs three key ingredients to keeping going: the first is heat (such as from the match used to light a candle), the second is fuel (the candle wax), and the third is oxygen (there's plenty of that in the air). Remove any one of those three things and the flame will die.

Dousing fire with water is effective because water is very good at removing heat and cutting off oxygen. However, our sun is not actually made of fire like a giant candle. It is a giant ball of plasma. Rather than combustion, the sun runs on a process called nuclear fusion, where the heat and pressure in the sun's core are so vast that small nuclei such as hydrogen are forced to fuse into larger nuclei such as helium, generating blistering amounts of energy and keeping us at a comfortable temperature here on Earth.

Enveloping the sun in a thick blanket of water wouldn't be much help if you wanted to snuff it out. Although the water would instantaneously remove some heat it would also increase the sun's mass, and therefore the pressure inside it, increasing the rate of nuclear fusion. What's more, the water molecules (consisting of hydrogen and oxygen) might get hot enough to be ripped apart into their constituent nuclei, providing more fusion fuel. So the sun would actually burn more fiercely and rapidly than before.

But what if you dumped all the water in the entire universe over the sun? Now you're talking. Strictly speaking, you'd be dumping ice rather than water, because space is so cold that almost all water exists in its solid form. You would theoretically be able to add so much to the sun's mass in the form of ice that it would use all its fuel very quickly. Then it would explode cataclysmically as a supernova, destroying Earth and leaving behind an extremely dense neutron star, or even a black hole. I guess you could count that as extinguishing the sun. In summary, getting the sun wet is likely to seriously mess up our solar system.

Sam Buckton
Hertfordshire, UK

Shape of things to come

? Comet 67P/Churyumov-Gerasimenko has an irregular shape. What size do comets or asteroids need to be before they develop an orb-like shape, and what are the processes involved?

Malvina Moray
Doune, Perthshire, UK

Calculations suggest that any rocky object with a diameter over 700 kilometres should be close to spherical. At this size the body has enough mass to overcome hydrostatic forces. Ceres, the largest object in the asteroid belt, has a diameter of approximately 945 kilometres and is nearly spherical. Because it is less rigid than rock, a body composed of water ice assumes this shape at much smaller sizes, becoming spherical at diameters of around 320 kilometres.

Assuming all planets have the same composition and density, these calculations also show that the maximum height of any protuberance – such as a mountain – should be inversely proportional to the diameter of the planet. This explains why Mars, with a diameter roughly half that of Earth, sports a mountain, Olympus Mons, that is more than twice the height of Everest. There is also the complication that bigger planets retain more of their primordial heat – which can be traced back to the gravitational potential energy of the gas cloud that collapsed to become the planet. This further reduces the height of mountains on bigger planets, as the rock underlying a mountain is warmer and less rigid.

Mike Follows
Sutton Coldfield, West Midlands, UK

Count the dims

? Presumably exoplanets can only be detected when the star they orbit dims as they pass in front of it as viewed from Earth. Have astronomers calculated what percentage of planetary systems meet this criteria? If so, how?

Paul Lea
Exeter, Devon, UK

Only a small percentage of exoplanets have been observed directly. The existence of others has to be inferred usually from observations of the host star. When a planet passes in front of its star as seen from Earth, the star appears to dim slightly. The measurement of these dips in brightness through a technique called transit photometry can yield a lot of information like the planet's orbital period and its size relative to its star. It also makes it possible to study its atmosphere and thus whether it might be home to extraterrestrials.

For an exoplanet to be observed in this way the planet, its star, and Earth all need to fall on the same plane. The probability of being able to detect such an alignment depends on the size of the star and the diameter of the planet's orbit. For a planet orbiting a sun-sized star at the same distance we are from the sun, the chance is around 1 in 200 that we would be able to detect it using transit photometry.

However, many other techniques can be used to hunt for extrasolar planets. One of the most successful is the radial velocity method. As a planet orbits, it will pull on its star, so the star is seen to move in a tight circle. Its light is blue-shifted while it is moving towards Earth, and red-shifted while it is moving away, and this can be detected using

Doppler spectroscopy. At the time of writing, nearly 2,000 exoplanets have been found, 90 per cent of them by one of the two methods described above.

Mike Follows
Sutton Coldfield, West Midlands, UK

Lunar lift

 Tides are affected by the moon's gravity. So does your weight change depending on its position?

Priyā-Louan Macknay, via email

Absolutely. The weight of any object anywhere in the universe is the sum of the attraction between it and every other body in the universe, according to their various masses and positions. This greatly complicates all sorts of measurements. For instance, in the eighteenth century the French astronomer Nicolas Louis de La Caille came to South Africa and made sophisticated measurements of the shape of Earth's southern hemisphere, concluding it was flattened. He did not realise that he should have made these measurements on a wide open plain. His various readings were badly distorted by the gravitational masses of Table Mountain and the Piketberg range, and it was some time before anyone corrected the shape he calculated.

As the moon passes above, to the west, below and to the east of us, we get lighter, lean to one side, get heavier, then lean to the other side. However, its mass is so far away that if I were to swing a pin on a silk thread, this would affect my apparent weight more strongly than the moon does.

Jon Richfield
Somerset West, South Africa

Long wave

[?] If a sea-level canal was dug from east to west across Asia, would the moon have a tidal effect on the water level, with a daily tidal bore in phase with it?

Lyndall Smith
Ferny Hills, Queensland, Australia

Tides are raised on Earth by the gravitational pull of the moon and, to a lesser extent, the sun. Even the water in your bathtub will be moved as the moon passes overhead, albeit by an immeasurably small amount. The same is true for water in a canal. But the tidal range – the difference in the water level at high and low tides – is only significant in huge bodies of water like oceans, and only noticeable when the bulge of water encounters a land mass.

Tidal bores occur in some rivers and, more rarely still, on lakes that are open to the sea, known as fjords in Norway or sea lochs in Scotland. A significant tidal range is required, typically in excess of six metres. Entering a wide bay, the incoming tide is funnelled towards the narrow opening of the river or lake, and the water piles up on top of itself, creating a wave that travels up the river. Bores can be sub-divided into hydraulic jumps, where there is a sudden change in water level, and undular bores. The UK's Severn bore is undular, characterised by a wavefront followed by a series of solitons, or solitary waves.

John Scott Russell first described solitons in 1834, when he saw one moving along the Union Canal in Scotland. It was created when a horse-drawn narrow boat suddenly stopped and the water that was being pushed along by the vessel continued for several kilometres, travelling at around 13 kilometres per hour. However, even if there were no locks, the energy of a tidal bore running along a vast trans-Asia canal would peter out as it worked against friction on the bottom and sides of the canal.

Mike Follows
Sutton Coldfield, West Midlands, UK

Fly me to the moon

? Is it possible to take the International Space Station out of its current orbit and land it on the moon?

David Anderson
Christchurch, New Zealand

Theoretically yes, but practically no. To move something from low Earth orbit to the orbit of the moon requires an increase in speed of about 3.2 kilometres per second. When it gets near the moon, it will be travelling faster than 2 kilometres per second relative to the moon – so this would have to be reduced to zero to allow a soft landing. The amount of conventional fuel needed for such accelerations and decelerations would be more than the mass of the International Space Station itself, and this fuel would have to be delivered to it from Earth or a source in space.

But the thing that makes it practically impossible is that you would have to bring the whole ISS down softly. The gravity on the moon is much less than on Earth, but this would still require attaching rockets all over the station and having them fire together. And even if it were possible to land the ISS on the moon without breaking it, it is not designed to sit on a surface, even under weak gravity.

Eric Kvaalen
Les Essarts-le-Roi, France

Hole number

? What happens when a black hole is swallowed by another black hole?

Jamie Malone
London, UK

It is not quite as exciting as you might think. The two black holes will fuse, creating a new one with a mass equal to the two constituent masses. The surface area of a black hole is related to its mass so this will increase in proportion. Because black holes are generally accompanied by an accretion disc of material that rotates around them moving faster and faster as it spirals inwards we can assume that the two discs will crash into one another before the holes meet. The end result will be a bigger disc but when the two very hot discs hit each other there will be an increase in the radiation given off until the new system settles down.

John Anderson
Warsaw, Poland

When two black holes approach one another they usually follow hyperbolic paths around a common focal point so they end up galloping off in different directions. If there are stars around, the situation becomes more complicated, and the black holes can end up orbiting one another after tossing some stars away. Such an orbit will continue almost forever unless there are still more stars nearby that can be thrown off, removing some energy and allowing the black holes to fall closer together. A black hole may also head straight towards another so that it would hit its event horizon. In that case, an observer would see the two black holes getting closer and closer (assuming they're visible), but would never see them coalesce. Instead, they would seem to slow down, get dimmer, and the light from them would get redder. This is because their time slows down so we might never see them reach the point at which one crosses the other's event horizon, even though that happens in due course.

Eric Kvaalen
Les Essarts-le-Roi, France

Galaxy twirl

? Are the spiral arms of galaxies evidence for gravitational waves? If not, what creates these arms?

Steve Miall
Higher Poynton, Cheshire, UK

Spiral arms are not evidence for gravitational waves. According to density wave theory, they are regions in which the density of gas and stars is much higher and this matter moves more slowly than in the regions between the arms. This theory explains how the spiral arms form in the first place – and once formed, they are self-sustaining.

Material that approaches a spiral arm speeds up because of the extra mass in the arm and then slows down while passing through it. This triggers intensive star formation, leading to the birth of massive and luminous stars that have relatively short lifespans (millions of years) compared with our sun (billions of years).

These types of star die in supernova explosions before they have time to leave the spiral arm. Only stars with a low mass, such as the sun, can leave the spiral arm and orbit the galactic centre to eventually return to their birthplace.

Chris Brindle
Bodmin, Cornwall, UK

3 Domestic Mysteries

My old china

? I have a china mug that has a vertical crack in it. When I drink cold liquids from the mug, it doesn't leak, but when I have a hot drink it does. I realise this is something to do with the china expanding with the heat, but surely it should expand into the crack as much as it expands in any other direction. So what is happening?

Marion Manders
Banstead, Surrey, UK

If a flat wall with a crack is constrained between two surfaces, then heating the wall will cause it to expand into the crack. However, an object like a china mug can expand freely in all directions. Hence, although the walls may be expanding towards each other in the tangential direction, they are also expanding away from each other in the outwards, radial direction.

Provided the heating and expansion are uniform, the relative proportions of all the lengths remain the same. The shape of the object – including any cracks – stays constant, it is merely enlarged. This enlargement means the crack is now wider, and so liquid can leak more easily.

At high temperatures, water has a lower viscosity so it can flow more easily through any cracks. It also has a lower surface tension at higher temperatures, which means it can wet a surface more easily. Both of these factors may be helping the liquid to flow out more readily.

Simon Iveson
New South Wales, Australia

If it warmed up very slowly, the whole mug would become slightly larger, retaining its proportions. This would make any existing crack slightly wider. But perhaps a better explanation lies in the 'differential expansion' that occurs as the mug is rapidly heated from the inside.

A curved wall will tend to straighten if warmed from the concave side. This is caused by the inner surface expanding first. So the wall of the cracked mug will try to unwrap and the crack will widen. Of course, the base of the mug restricts this effect – especially if it is relatively thick and so warms more slowly. However, the crack is still there, it will just be wider at its top, essentially a very thin V shape.

Martin Wilson
Kendal, Cumbria, UK

Terracotta solder

? I put a piece of damp terracotta into my brown sugar to keep it crumbly. Why does the same treatment make icing sugar fuse into a solid brick?

Kerry Graf
Florey, Australian Capital Territory

Brown sugar is a mix of crystalline sugar and up to 10 per cent molasses, a brown syrup obtained during the sugar refining process. Soft brown sugar is moist due to the hygro-scopic (water-absorbing) properties of the molasses. If this water is allowed to evaporate, the sugar will harden. To prevent this, you can add a piece of damp terracotta, a slice of bread or a few marshmallows to supplement the moisture.

Icing sugar is an entirely different kettle of fish. It is a finely ground sugar that, due to its small particles, is prone to caking or clumping if it gets moist. Such highly refined, granulated sugars have a water content of around 0.05 per cent and contain anti-caking agents such as calcium phosphate to absorb moisture and prevent clumping. The correspondent's damp terracotta overcame the anti-caking agent, so the sugar particles clumped then crystallised together.

Experienced bakers know that success with cakes and biscuits depends on using the right sugar for each job. That's the way the cookie crumbles.

David Muir
Science Department, Portobello High School, Edinburgh, UK

Boil foils

? I reheated a mug of instant coffee in the microwave but left it too long and it was boiling over when I took it out. I ran a little cold water into the mug to cool the coffee and it boiled again. The tap water ran in gently and large active bubbles appeared on top of the coffee. How could this happen?

John Davies
Lancaster, UK

The coffee was probably not actually boiling; rather dissolved air was escaping. Take a pan of tap water, put it on the stove and start heating it. Very quickly tiny bubbles will form on the surface even when the water is quite cool. You can test this (very cautiously) by dipping your finger in. Tap water is often saturated with air. With most gases the solubility in water decreases as the temperature rises so the dissolved air escapes when the water is heated.

The air bubbles expand as the temperature rises and they gradually detach from the pan's sides and rise quickly to the surface of the water where they burst. What has any of this to do with the coffee miraculously appearing to come back to the boil? When the cold tap water was added to the cup the dissolved air in this water would come out of solution very quickly as it heated. As a result these bubbles would rapidly expand and burst. It is very difficult to visually distinguish between air bubbles forming, expanding and bursting and steam bubbles doing the same – but only the latter counts as boiling.

Peter Borrows
Amersham, Buckinghamshire, UK

Boiling depends on bubbles of water vapour being able to form within the liquid that is being heated. That process is helped by the presence of nucleation sites – for example sharp points on surfaces, suspended particles and existing small bubbles – all of which help new bubbles to form and grow. If the pressure inside a bubble is less than that of the atmosphere on the liquid's surface the bubble collapses. It is the noise of collapsing bubbles that makes a kettle 'sing' just before it comes to the boil.

Once the liquid has reached its boiling point, the vapour pressure inside the bubbles is sufficient for them to persist and rise to the surface, carrying molecules of the liquid into the air. This is where things get less straightforward. The pressure required to stop a bubble of vapour collapsing is actually slightly greater than that on the liquid's surface. Perhaps surprisingly, the excess pressure is inversely related to the bubble's radius. This means that the smaller the bubble, the higher the pressure needed, and since pressure rises with temperature, so the hotter the bubble needs to be to persist. This helps to explain why a liquid can be heated well above boiling point without boiling, provided no bubbles or sharp edges exist inside.

Such a liquid is said to be superheated. Microwaves heat a cup of coffee unevenly. So even though the coffee was boiling over when taken out of the microwave, parts of it may remain superheated after the existing bubbles have escaped. Indeed, the average temperature may stay above boiling point. The cold water that is then added to the mug immediately warms and gives up small bubbles of air (air is less soluble in warm than in cold water). These bubbles bring the coffee to the boil once again by allowing new bubbles to form within the superheated regions.

The coffee can now boil vigorously which is potentially dangerous, as spitting and foaming of the hot liquid may occur. This is why microwave ovens and microwaveable meals often carry warnings to let food stand for a minute or so after cooking before being moved or stirred.

Alan Goodwin
Cheadle Hulme, Cheshire, UK

Put a lid on it

[?] I know high pressure allows us to boil vegetables more quickly, and this is obvious when a pan lid is fitted tightly. But I can't figure out why the pressure is still increased when the lid only partially covers the pan, leaving a 1 or 2-centimetre gap. It seems to me that air and steam can pass easily through this space and circulate freely, yet even a partially covered pan of liquid seems to boil more quickly. Why?

Peter O'Regan
Dublin, Ireland

Partially covering the pan disrupts key mechanisms by which boiling water loses heat, namely evaporation and convection. A lid, close-fitting or otherwise, will reduce the convective cooling by preventing fresh cooler air from accessing the heated liquid.

The same applies to the evaporated water, which will condense almost immediately on the underside of the lid, returning much of the latent heat of evaporation back to the liquid in the pot. However, this is not the complete answer, because an old proverb points to the quantum features of the pot: if continuously observed, we know it will never boil.

Neil Barry, via email

I have always assumed that the rapid boiling of water in a pan covered partly by a lid is the result of heat retention, rather than marginally increased pressure.

The small aperture left by a loosely covering saucepan lid will mean less heat loss, therefore the energy in the partially closed system can climb higher. This may produce a stable but more active boiling liquid than when the lid is off. The alternative is that 'too much' energy is retained and the simmering liquid becomes over stimulated, resulting in a rapid, catastrophic loss: the dreaded boil over.

Clive Tiney
York, UK

Water boils faster with a lid on the saucepan, but it is not because of increased pressure. The lid allows steam to condense on its surface and returns it to the pan, along with the heat it would otherwise carry away. This will be most effective with the lid fully on, but will still work with the lid only partly covering the pan. The more steam that escapes, the greater the heat lost.

Guy Cox
University of Sydney, Australia

Nice rice?

? One often hears that it's not safe to eat reheated rice. Why is this? Surely anything that is living in the uncooked rice is killed by boiling it. If you then freeze the rice very quickly it must be pretty much sterile and can be reheated when required. I've done this for years and am unaware of any ill effects. Am I dicing with danger?

Ellen Finnan
Aberdeen, UK

Bacillus cereus is the most common cause of poisoning from reheated rice. It is found in soil and the heat-resistant spores can survive boiling, allowing them to germinate when the rice cools. Fortunately, it seldom invades your organs, but infected rice toxins produced by the bacteria may cause nausea and a painful case of diarrhoea within hours of eating.

Healthy people generally recover in a day or two with no more treatment than staying close to a lavatory and drinking plenty of rehydrating fluids, but it may be difficult to tell an acute attack from more dangerous types of food poisoning. Some strains of *Bacillus cereus* may also themselves be dangerous, so don't complacently refuse to see a doctor if necessary.

These effects are largely avoidable. To forestall any problems, cook your rice until it's hot and eat it soon. If you want to keep any, prevent the bacteria from multiplying by either keeping the rice too hot to touch (60°C or higher) or cooling it to fridge temperature (about 4°C or lower) as fast as possible, even if that means putting it warm into your fridge or freezer. Such modest heat and fast cooling won't kill spores, but will prevent them from germinating and producing toxins.

Jon Richfield
Somerset West, South Africa

Gloomy-side up

Whenever I make scrambled eggs they turn out grey. What causes this?

Les Moore
Hartford, Cambridgeshire, UK

If you remember your chemistry teacher heating a mixture of iron filings and powdered sulphur to form iron sulphide, with a whiff of eggy hydrogen sulphide in the air, you've experienced the chemistry behind the grey eggs. Iron from the yolk combines with sulphur from the egg white to form grey iron sulphide, especially so when overcooked. If you cut a hard-boiled egg in half you will see a band of greenish-grey iron sulphide at the interface of the yolk and egg white. The colour is caused by the mixing of the grey sulphide and yellow yolk.

To avoid grey eggs cook them less, or if you insist on having well-cooked eggs, add a drop of yellow food colouring to brighten up breakfast.

David Muir
Science Department, Portobello High School, Edinburgh, UK

Hisssssss

? How long do I have to wait after dropping a full beer bottle before I can safely present it to a guest, and how did its internal pressure build up if the system is in equilibrium?

Robert Volpe
Highton, Victoria, Australia

The pressure inside a bottle of beer is unchanged after it has been dropped. What has changed is the number of tiny bubbles that are distributed throughout the beer. These form when a bottle is dropped or shaken. When the bottle is opened the bubbles act as nucleation points for the dissolved carbon dioxide which comes out with such a rush that it takes a lot of the beer with it. This is a terrible waste: not only is good beer lost but what remains is flat. As the writer suggests, this doesn't happen if you wait before opening the bottle. How long you must wait depends on a number of things: the size of the bubbles, how deep in the liquid they are, and the viscosity of the beer. To be safe, one has to wait long enough for the last smallest bubble to reach the surface and burst before removing the cap. In my experience, about ten minutes is normally enough. However, a ten-minute wait may be asking too much of a thirsty beer drinker, so it is far better to place the bottle at the back of the refrigerator and take out another – with greater care.

Andrew Carruthers
Quebec, Canada

Foaming mad

? I have a hand-held milk foamer for my coffee. I can foam milk to twice its volume when it is cold out of the fridge, but it barely foams at all when heated. Why is that?

Mark Alberstat
Halifax, Nova Scotia, Canada

Creating a milk foam depends largely on the protein casein, which stabilises bubbles in the milk. The protein forms particles called micelles, and the stronger the structure of these particles, the greater the volume and durability of the foam.

Milk also contains many different fats in the form of long fatty acid chains attached to a glycerol backbone. Heating it increases the activity of an enzyme called lipase. This encourages lipolysis, a process by which fats break down into their constituent fatty acids and glycerol. The glycerol and free fatty acids can weaken casein's structure, causing air bubbles to collapse more easily.

If the milk has just been removed from the fridge, its fats are largely intact, so casein is able to stabilise any foam. The freshness of the milk also has an effect on foaming, because the longer the milk is left to stand, the more of its fats will be broken down. To ensure a good froth, it helps to use milk with a low fat content such as skimmed milk, so there are fewer fat breakdown products to weaken the casein. While this should help create a fuller foam, the taste may leave something to be desired.

Sam Buckton
Hertfordshire, UK

Leather and steel

? As a boy I watched my grandfather sharpen his cut-throat razor on a leather strap. How did this work?

Phill Wells
Windsor, Berkshire, UK

Stropping a blade on a leather strap (or strop) as described is supposed to smooth out indentations without removing any metal. In contrast, sharpening a blade on a whetstone removes any metal bent out of alignment. The leather is usually impregnated with an abrasive compound like jeweller's rouge or green chromium (III) oxide. To avoid cutting or nicking the strop while achieving a sharp edge, the correct technique is required: the razor has to be held flat to the leather and drawn spine first along the length of the strop.

Mike Follows
Sutton Coldfield, West Midlands, UK

Bowled over

? Water moves about in my toilet bowl when it's windy outside. Why?

Tony Sandy
Kilmarnock, Ayrshire, UK

The plumbing of domestic drains is behind this phenomenon. Your toilet bowl has a continuous connection to the sewer underground, but it also contains a connection to the open air. The purpose of this vent pipe is to ventilate the soil pipe which carries water and waste material from the toilet to the sewer. Your toilet has a U-bend which holds water to form a seal and prevent noxious fumes drifting from the sewer into the bathroom. When the toilet is flushed, water rushing down the pipe creates negative pressure behind it which would drag this water seal out of the U-bend. The vent pipe prevents this by allowing some outside air to be pulled into the pipe instead.

The vent pipe needs to be installed with its outlet away from windows because its direct connection to the sewer will allow gases to escape. This is why most are at roof level. When the wind blows over the end of the vent pipe, the Bernoulli effect comes into play, creating suction in the pipe which draws water out of the U-bend. If the wind is gusting, water in the toilet bowl can move up and down.

Malcolm Nickolls
Aylesbury, Buckinghamshire, UK

Waxwork horrors

? I left a red tea-light candle on my windowsill for six months in a transparent plastic box. It now looks very strange, with pale 'roots' growing within the wax. How did this happen?

Rose Howie, via email

I believe the pattern has formed due to the crystallisation of paraffin molecules, perhaps caused by daily warming on the window ledge. Paraffin has a tendency to dissolve as the temperature rises, followed by a tendency to crystallise as the temperature falls. This is called thermally cycled recrystallisation, and is commonly used in industrial processes with sugar, salts and so forth. It is also used to form crystals out of biological proteins to allow for X-ray diffraction analysis of a protein's molecular structure.

A clue to this crystallisation is the lack of colour in the 'root' growths. Because the pigment in the candle wax is a different molecular shape and size to the paraffin molecules, the paraffin molecules displace the dye as the crystals grow and shrink. A close examination might show a coloured halo around the crystals where the expelled dye is a little more concentrated. Eventually the paraffin molecules will join together to become a mass of pure paraffin.

Bill Jackson
Toronto, Ontario, Canada

Liquid gold

? I recently saw vodka in the supermarket with real gold leaf flakes in it. What caught my eye was that the gold flakes were distributed uniformly throughout the bottle. Gold is much denser than alcohol, so why don't they sink to the bottom?

Andrew Menadue, via email

Left long enough, the gold flakes will fall to the bottom of the bottle. The reason they remain fairly evenly distributed is that they are small and light enough to keep floating. Gold is incredibly malleable, meaning it can easily be worked into flakes a thousandth of a millimetre thick.

Energy from the bottles being moved around periodically, or from floor vibrations as people walk past, and convection currents from temperature changes in the liquid are enough to disturb the flakes and set them floating. The same effect applies to dust, soot and smoke particles in the air.

Ian Gordon
Camberley, Surrey, UK

One product that displays the behaviour described is Smirnoff Gold Cinnamon Flavoured Liqueur. If a bottle of this is rotated gently through about 45 degrees clockwise, the gold flakes follow the clockwise motion, but will then 'rebound' anti-clockwise when you stop. You can often observe similar rebounding if you stir a bowl of soup. This is not the behaviour of a Newtonian fluid, so the bottle probably contains something other than the usual constituents – perhaps some gelling agent.

The gelatinous motion of the bottle's contents is more reminiscent of a weak Bingham plastic, a material that remains rigid until it experiences a sufficient level of shear stress to start behaving as a viscous fluid (for example, mayonnaise).

This can be confirmed by shaking the bottle and observing the motion of the air bubbles: as they rise, you can clearly see that they are threading their way through an invisible solid matrix, and some bubbles even become trapped in it.

Terry Collins
Harrogate, North Yorkshire, UK

Clean break

? I've discovered that glasses with a thick base crack around the bottom when cleaned in the dishwasher. Why is this?

Tony Sandy
Kilmarnock, East Ayrshire, UK

When I was a child helping wash dishes, my mother would tell me not to rinse hot glasses with cold water because they might break. I later learned that glass has poor ductility, so stress created by different thicknesses of the material contracting by different amounts can cause it to fracture.

Glass, like many substances, expands and contracts with temperature. Glasses of variable thickness are more likely to break with thermal shock because the thin parts will heat or cool faster and thus expand or contract faster than thicker parts. Dishwashers use water at a much higher temperature (between 70° and 75°C) than you would use to hand-wash dishes (between 40° and 45°C). When the hot water in your dishwasher hits a cold, thick-bottomed glass, it causes differential expansion, resulting in a crack.

Some people use hot water to defrost car windscreens. For the same reason, they may have found it to be a shattering experience. Older windscreens are more likely to break because chips and scratches act as nucleation points for fracturing.

David Muir
Science Department, Portobello High School, Edinburgh, UK

Everything but the curl

? I've been making notes to revise for my exams, and as I write the pages curl inwards first from the top corners of the paper, and then the bottom corners. Why is this? I'm using a black ballpoint pen on white A4 printer paper.

Gurleen Kaur, via email

I suspect your note-writing is quite small and closely spaced, and that you apply more pressure than normal with the pen. This compresses the fibres of the paper and causes them to expand sideways, thus effectively stretching the paper below the pen. Because you don't write to the top, bottom, and edges of the sheet, these parts remain unaffected and so curl up.

A classic example of this effect but for a different reason is a sandwich left out in the sun. Most people will notice that the corners and edges soon start to curl and it will become unpalatable. This time the phenomenon is caused by the sides and upper surface of the bread drying out and shrinking.

The stretching effect caused by pressure is used by manufacturers to produce curved skin panels for custom-built cars and aircraft. A machine called an English wheel is used for this purpose. This consists of a rigid 'G'-shaped frame supporting a large flat wheel. Below this is a smaller wheel with a curved edge. A sheet of metal is fed between the wheels and worked back and forth. In the hands of a skilled operator, quite complex shapes can be produced. There are many videos online that show the machine in action. Before the invention of the English wheel, the stretch-shaping of metal was achieved with a hammer.

Bob Halahan
Godalming, Surrey, UK

Fur-up ball

? I live in a hard-water area and use a stainless-steel mesh ball in my kettle to stop it 'furring up'. The steel mesh traps the scale that would otherwise form inside my kettle. How does it work?

V. M. Koritsas
Otley, West Yorkshire, UK

If you want to get rid of something, you can use a chemical, physical or biological agent – or you can provide a different environment that it prefers. Some gardeners use sacrificial planting of nasturtiums to attract aphids, and thus protect their prized plants rather than using pesticides. Aphids can then be physically removed from the nasturtiums or the most infected plant parts can be destroyed.

In a similar manner, limescale is deposited preferentially on the stainless-steel mesh ball because the mesh has far more nucleation sites than the kettle. These sites are points such as microscopic scratches where the crystallisation of limescale (mostly calcium carbonate) starts. The mesh ball won't capture all the deposited limescale, but it can be periodically soaked in vinegar to dissolve the build-up and then reused. Don't be tempted to use cheap wire wool bought in hardware stores. This will corrode and the rust will make your tea and coffee browner than usual.

David Muir
Science Department, Portobello High School, Edinburgh, UK

Laundry stiffener

? To do our bit for the planet, we have been hanging our washing on a line in the back garden rather than using our electric tumble dryer. Thicker materials such as bath towels and socks are hard and abrasive when dried on the line, in comparison to the soft and fluffy feel of the tumble-dried versions. Why does this happen?

Ronan Tierney
Enniscorthy, Ireland

What is happening to the heavier fabrics in the tumble dryer is 'felting'. This is due in part to the centrifugal force applied by the action of the dryer, and also in part by the capillary attraction between fibres as the water drains away. When air-drying, if the water items are washed in is hard, then the minerals it contains will precipitate out, which helps to cement the fibres together.

To counteract this, you have to simulate the action of the tumble dryer in separating the fibres before drying takes place. The real experts in this are sea otters (though they don't have to deal with spin or tumble dryers). Watching video of them fluffing up their own fur and that of their cubs might be instructive – or at least entertaining, with a high 'ahh' factor.

As a single person with ready access to outdoor drying, I cannot justify having a tumble dryer. Every towel gets vigorously shaken before being hung out to dry. Bath mats are a bit too heavy, so their nap is fluffed up by brushing with an outstretched hand. It works for me.

Mike Coon
Maidenhead, Berkshire, UK

Made to last

[?] I have a garden table and four chairs that I bought about fifteen years ago. I leave them outside, uncovered all year round in sun, rain and snow, yet they still look as good now as they did the day I bought them. I want to know what they are made of and what kind of paint is on them.

They are metal and relatively lightweight. On a couple of spots where the metal has been exposed through the paint, it is a shiny silver colour which never rusts. The paint itself is light blue and it never flakes. The only way to damage the paint is to really hit it hard, and then a bit of the metal may be exposed. The set really is remarkable, and I wish I could remember where I purchased it.

Diana Smith
Oxford, UK

The correspondent's garden furniture is likely to be made of aluminium, a low-density metal that resists corrosion. When metals such as iron rust, the resulting iron oxide flakes off, exposing fresh metal to further oxidation. Unless protected and preserved in some way, iron structures will tend to eventually corrode right through and become physically weakened. This is why checking for corrosion is part of the UK's MOT roadworthiness test for motor vehicles.

In contrast, when the surface of aluminium oxidises, it stays where it is, forming a dull protective layer – although shiny metal is exposed when this is scratched off.

Aluminium needs to be prepared with an etching primer so that paint will adhere to it. Tough polyester powder paint might have been applied but, if destined for outdoor use, a latex paint would be ideal as it expands and contracts with the big changes in temperature that outdoor furniture will experience.

Mike Follows
Sutton Coldfield, West Midlands, UK

The furniture is probably made of aluminium, a light metal that develops a protective layer of aluminium oxide when exposed to air. The oxide layer prevents further corrosion because it is impervious to air and water and doesn't flake.

The layer can be strengthened by what is known as anodising. In effect the aluminium object acts as the anode in an electrolytic cell and oxygen reacts with it to further thicken the oxide layer. Another advantage is that the oxide layer can be dyed, so the colour becomes an integral part of the surface structure. This may be the case with this garden furniture. In this way, aluminium objects can be given attractive, coloured finishes that can withstand weather and children.

David Muir
Science Department, Portobello High School, Edinburgh, UK

Mystery measure

?   Why do some imperial tape measures have a mark at 16, 32 and 48 inches, and so on?

John Jarvis
Bristol, UK

The extra marks found at 16-inch intervals are simply the standard spacing for joists and studs in carpentry. Joists are the horizontal beams that support wooden floors, and studs are the wooden bars that make up the frame for a hollow, non-load-bearing wall that is faced with plasterboard or other similar sheet material. Having the intervals marked like this saves the carpenter from having to know the 16 times table.

Prompted by this question, I checked the spacings of the joists in the floors of my Victorian house, and they are at about 18 inches. Perhaps the Victorians could afford thicker floorboards that needed less support. Interestingly, I have a metric-only tape measure which doesn't have marks at 40-and-a-bit centimetres (the metric equivalent of 16 inches).

David Jackson
Liverpool, UK

4 Human Body

Heart to heart

?　Why do animals have organs? Surely it would be more efficient and robust for us to have a thousand tiny hearts distributed round our bodies rather than one big one in the middle?

Adrian Bowyer
Foxham, Wiltshire, UK

Insects and other small creatures have no hearts; instead they have a dorsal vessel that contracts in several segments to circulate haemolymph around the body. This fluid does not carry oxygen, however, because the gas exchange required for respiration can be accomplished by simple diffusion from the surrounding air. Bigger creatures such as ourselves need a more densely packed surface for gas exchange (such as gills found in fish or lungs in other animals), a network of blood vessels to transport the gas, and a pump to push the blood around.

The heart is evolution's best attempt at a high-pressure pump, and in terms of power-to-weight ratio it is likely to outperform a distributed system. However, this arrangement does introduce friction between the blood and the vessel walls. Therefore the blood pressure needed to operate a closed circulatory system increases with body mass. Taller animals must also force blood higher against the force of gravity.

The idea of having redundant hearts is nice, but for this to work each little pump would need to have its valves open if it fails, so that blood could still be pushed through it. To maximise efficiency each heart would have to beat when the pressure at its inlet was at a maximum, a far more complex task to manage than a single heartbeat.

Mike Follows
Sutton Coldfield, West Midlands, UK

Night vision

? If I keep my eyes closed when in a dark room, I start to see lights. They occur in two forms: bright sparks like stars in a night sky and, after a while longer, irregular patches of blue/purple that coalesce into rings that get progressively smaller until they disappear, only to be replaced by another similar patch. The patterns get more complex as time goes on. The lights disappear when I open my eyes and admit the faint light of the room. Are these experiences universal and how do we explain them?

Chris Gilfillan
Surrey Hills, Victoria, Australia

This phenomenon is called a phosphene and arises because our visual system never shuts down. Even in the absence of photons, and when our eyes are closed or when we are asleep, the neurons in our visual system are constantly firing. 'Seeing stars' can also be caused by a blow to the head or other mechanical stimulation, like rubbing our closed eyes when tired.

Isaac Newton risked blinding himself by taking his investigation into vision a step further. His notebooks (*bit.ly/*

NewtonNotes) reveal he distorted the shape of one of his eyeballs by poking a bodkin between it and his eye socket, and noted the patterns he observed. Patterns can also be induced by the drop in blood pressure experienced when we stand up too quickly, or by consuming psychedelic drugs such as LSD or psilocybin.

Perhaps the question should be turned on its head. Why should it be dark when we close our eyes? After all, infrared radiation (heat) from the eye typically accounts for about a million times more energy than visible light streaming through our pupils when our eyes are open. It turns out that infrared photons do not have sufficient electromagnetic energy to be detected at the retina.

Mike Follows
Sutton Coldfield, West Midlands, UK

ZZZZZZCHOOO . . .

? Why don't we sneeze in our sleep?

Tony Sandy
Kilmarnock, Ayrshire, UK

It is a leading question, implying that 'we' means all of us. I do sneeze during my sleep, frequently if I have a head cold. This wakes up my wife who then awakens me and tells me that I have disturbed her sleep by sneezing yet again. Presumably, if I slept by myself I would not have woken up. And the implication of this is that only couples are known to sneeze while they are asleep. Now there is a PhD project.

Ron Pursell
Flookburgh, Cumbria, UK

A sneeze is a reflex reaction – an involuntary response to an internal or external stimulus. Reflex reactions shut down during sleep because of rapid eye movement (REM) atonia, a physiological state during which movement messages can't get from the brain to the muscles, causing partial paralysis. This prevents sleepers from acting out their dreams to the detriment of themselves or their bed partners. A consequence is that reflex signals are not acted upon, so there's no sneezing during sleep. If a sneeze is absolutely necessary, REM atonia ceases and the person wakes up before sneezing. There have been cases of REM sleep behaviour disorder in which people acted out their dreams and ended up in court after injuring or killing their partner. Presumably such people can sneeze in their sleep.

David Muir
Science Department, Portobello High School, Edinburgh, UK

Crash, bang, wallop . . .

? Why do loud noises seem to hurt? They don't seem to cause physical pain but do induce mental anguish. I heard a truck taking on board a bin of empty bottles and the noise was excruciating. I had to get out of earshot.

Alan Vatt
Lancaster, UK

Two factors contribute to the unpleasantness of the noise from bottles going into a truck: the volume and the discordant frequencies. The human ear detects sound using tiny hair cells in the cochlea. A sound wave causes mechanical deformation of these cells sending an electrical signal to the brain via the auditory nerve. Different groups of hair cells react to different frequencies, with hairs detecting low notes at one end of the cochlear spiral, and high notes at the other. High volumes can damage the delicate hearing apparatus from chemical exhaustion of the cell-signalling process and mechanical damage.

We are adapted to find damaging stimuli unpleasant and painful so that we avoid them. Humans also find notes that are too close together very unpleasant – think of the discord between two adjacent notes on a piano compared with reasonably pleasant sounds obtained by playing notes spaced apart. This is probably because the closeness of the frequencies means there is an overlap in the groups of hairs activated by each note overstimulating those that can sense both. Glass colliding as it is dumped into the truck creates many discords at once. For high notes like the crash of breaking glass the problem is worse because as the notes get higher the difference between adjacent notes gets smaller and more hairs will be stimulated by a given note. We are therefore more likely to experience discordance at higher frequencies.

Miriam Ashwell
London, UK

Rather than ear damage the issue here could be hyperacusis – an oversensitivity to certain types of sound. Some of us might have experienced this during a hangover when even tiny noises seem dreadful. Hyperacusis can occur on its own or as part of a number of medical conditions. As a doctor I see this most often in migraines. Having said this, the questioner probably doesn't have true hyperacusis. His account sounds more like he has an aversion to that one particular sound. Now my kids are growing up it's wailing babies that set my nerves on edge, giving me the greatest urge to run for the hills.

Robert Ewing
Edinburgh, UK

Sense of proportion

? Why do we care so much about restoring lost sight and hearing, but give relatively little focus to restoring taste, smell and touch?

Fred Appleby
London, UK

It is only natural that we care most about the senses that are critical for our immediate survival. Arguably, sight is the most useful sense. It allows us to move at high speed through our environment, while avoiding hazards like low-hanging tree branches and cliff edges. It also allows us to notice problems and opportunities to find food, as well as greatly helping us to fend off predators.

Although blind people can learn to compensate for lack of sight, they do so within a community of people who are sighted. If all humans were to become suddenly blind, then our modern transport system would literally come crashing to a halt and civilisation as we know it would collapse – as happened in John Wyndham's science fiction novel *The Day of the Triffids*.

Hearing is also very useful. It facilitates teamwork and communication of knowledge that may be important for our survival. Those who become deaf can easily become socially isolated. Of course, sign language and the written word can also be used to communicate, but these only work when the recipient is looking. In contrast, we can hear and respond to warnings shouted to us from any direction.

Unlike blindness, if the world's population were to become deaf overnight, it is quite conceivable that civilisation could survive, albeit without a radio or music industry.

The senses of touch, taste and smell are less immediately essential for day-to-day survival. Obviously, taste and smell can warn us if food is not fit to eat, but these senses are poorly developed in humans – I have on occasion unwittingly eaten spoiled food that later caused me illness. Similarly, the loss of the sense of touch makes fine motor tasks such as doing up buttons and cracking an egg more difficult, but not impossible, provided we can see what we are doing.

It should be noted, however, that eating loses its rich pleasure without a sense of taste and smell. There is much research to suggest that in the long term this results in much higher rates of depression in people who have lost these two senses.

And without a sense of touch, the cumulative effect of not noticing all those bumps, burns and scratches results in serious skin infection, as seen in people with leprosy.

Simon Iveson
Callaghan, New South Wales, Australia

Footloose

I sometimes notice that one of my shoes is loose although the other is fine. When I tighten the laces on the loose shoe, the other shoe then feels loose. Is there an explanation for this peculiar effect?

Andrew Brittain
Sutton, Surrey, UK

Our senses are slaves to relative comparison. Place your left hand in hot water and your right in iced water and then put both into the same tepid water. Now your left hand senses cold and your right feels warm, despite experiencing the same temperature. When we walk from a sunlit area into a dimly illuminated room we see next to nothing initially, yet within seconds our eyes compensate and we can discern subtle decor and different shades of grey. When loud raucous music is switched off silence prevails for a moment until we perceive hitherto unnoticed sounds of far-off traffic or birdsong. It can be irritating when the laces on one shoe loosen so we retie them tight. The other then feels loose because by comparison it really is looser.

David Muir
Science Department, Portobello High School, Edinburgh, UK

Head lines

? I have noticed that my head is tilted to my right in most photos even though at the time the photos were taken I believed my head to be erect. If I move my head to be upright I feel as though it is tilted to the left. I am strongly right-handed. What causes this delusion and is it common?

Alan Bartlett
Mansfield, Queensland, Australia

There may be an anatomical explanation for your questioner's head tilt. Our son was diagnosed at age thirteen as having a rare anomaly of his C1 vertebra: it is open in the front and has a floating piece of bone instead of part of the left side. The anomaly was discovered in a neck X-ray while neurologists looked for the cause of two strokes he had.

They suspected the problem was in his neck because prior to the discovery physical therapists often coached him to hold his head straight. Photos of him as a baby and young child reinforced their hunch: he had always held his head tilted to the left. The paediatric cervical spine specialist later confirmed that the C1 anomaly would cause a tip in the angle at which he holds his head. The anomaly had indirectly caused his strokes by repetitively injuring his left vertebral artery. He has made a full recovery through time, hard work and a titanium microcoil implanted to protect the affected artery. Your correspondent's experience may be 'a delusion' but evaluation by a spine specialist might be recommended.

Nancy Benz
Ramsey, Minnesota, US

As an occupational therapist, I have seen many people who have had strokes subsequently lean their head and even their entire body to their unaffected stronger side. Many people who have not had strokes also have a tendency to lean a little. There are two factors at play here. First, the brain adapts to conditions. After leaning slightly for a period of time the brain accepts this condition as normal and perceives the body to be upright. Second, very right-dominant people use their right arm and hand more than their left so they develop strength and muscle tension more on that side. Increased strength and tension in the upper trapezius and levator scapulae muscles which connect the shoulder girdle to the head can cause the head to tilt to the right and the brain would soon adapt to accept this position as upright.

Bonnie Clancy
Fort Myers, Florida, US

Mind lags body

? I'm seventy-seven and like many people my age tend to stoop rather than stand up straight. A few months ago I was standing by the kitchen door talking idly. For some reason I turned and hit my head a stunning blow on the door frame. More recently I was looking over my wife's shoulder as she worked on her computer. I reached out my hand pointing to something on the screen. As I pulled my hand back my fingers got caught in a box on the table and I toppled it over onto the floor spilling its contents. After a while I realised that both times I'd been stooping. In my mind's eye my head was clear of the door frame and my hand well above the table but my stoop cancelled both. Is our body image slow to adjust to reality? If so, why and can it be fixed?

Peter Laurie
Abbotsbury, Dorset, UK

The subconscious perception of your body orientation and movement is called proprioception. This aspect of awareness is supplied by the vestibular system in the inner ear, and by nerves in the muscles and joints. It lets you know where your hands and feet are when you are blindfolded.

Proprioception is learned through experience but can be compromised by changes in body structure due to adolescence, injury or ageing. Teenagers who go through a growth spurt may, when playing sports, start to drop balls that they used to catch; their arms are now longer than their brain remembers. The adolescents' proprioception lags behind their growing bodies.

However, neuromuscular pathways can be retrained with specific practice: evidence shows that juggling helps cricketers. The correspondent's proprioceptive experience is based on more than seventy years of standing tall. This conflicts with his current stoop and results in accidents. The good news is that movement and awareness exercise such as t'ai chi or yoga not only improves proprioception, but diminishes the likelihood of the falls and broken bones that are the bane of older people.

David Muir
Science Department, Portobello High School, Edinburgh, UK

I read with great interest your earlier correspondent's explanation of how proprioception may fail to keep up with changes in body shape. Four years ago I was pregnant with twins. Even though I could see my quite sizeable belly, touch it with my hands, and sense the twins kicking my internal organs I couldn't feel my belly 'from within'. When lying in bed on my back with my eyes closed, I could feel the blanket touching my belly but my abdomen seemed to me to be flat and in the very same place it had been all the years previously. I recall having to be especially careful when shutting the car door: if I didn't make a conscious effort I'd slam it on my bump. Apparently nine months were not enough for me to develop a proper proprioceptive feeling for my new body shape.

Ilka Flegel
Kapellendorf, Germany

Disease-proof

? If you have a few drinks, does the alcohol in your bloodstream act as medication against any microbes in the body?

Adam Wilson
Leigh-on-Sea, Essex, UK

There can be a protective effect, but only if the alcohol concentration is high enough. A study in the journal *Epidemiology* is one of the few to confirm the benefits of drinking alcohol with a meal. During an oyster-borne outbreak of hepatitis A, those drinking alcohol of 10 per cent strength or higher experienced a protective effect – although this would have occurred in the stomach, not the bloodstream.

I would argue that for the sake of your health, total daily consumption should be moderate, the equivalent of two glasses of table wine.

Lewis Perdue
Sonoma, California, US

A labour lost?

? While perusing my Shakespeare this evening, I wondered if it would be possible for anyone to accurately memorise the complete works – even if only by rote? Given that prodigious feats of memory were commonplace before we all learned to read and subsequently lost the ability to remember whole sagas at a sitting, would even the highly trained and practised

balladeers of the time have had the ability to accomplish
this feat (or its equivalent)?

Doug Hunt
Auckland, New Zealand

Memorising a large body of work requires interest, practice
and time. There are about 880,000 words in the complete
works of Shakespeare. To put that into perspective, the Bible
has around 750,000 words, and it started life as an oral
tradition.

It was said of George Fox, founder of the Quakers in the
1600s, that if all the Bibles in the world were lost, this would
not be a problem, because he could recite it completely.

I expect some members of the Royal Shakespeare Company
will know many if not all of Shakespeare's works by heart.
For most of us, however, access to all the knowledge in the
cloud through our smartphones and computers means we
only need to know how to find our favourite search engine
through the latest operating system. So these feats of memory
will become rarer and less believable as technology progresses.

John Wood
Winster, Derbyshire, UK

Resistance not futile?

Do people who have recovered from regular
tuberculosis have greater immunity to multi-
drug-resistant TB?

Eileen Caiger Gray
Doncaster, South Yorkshire, UK

Surprisingly, given the historic nature and the incidence of tuberculosis in some parts of the world, the actual TB bacillus isn't particularly infectious. It is thought that only 10 per cent of people who are infected develop active TB. The vast majority of people exposed to or infected by the bacillus will never go on to develop any clinical illness. For the most part you need to be exposed repeatedly, in for example close family units or bad housing, with malnutrition and poor immunity possibly playing a part.

In the pre-antibiotic era, a substantial proportion of people with active TB eventually recovered. Even today a small subset of people with multi-drug-resistant TB will have apparent clinical recovery. So both innate resistance and acquired immunity against tuberculosis seem to exist. Does a person who has been cured of TB have greater resistance to a new external source (which may or may not be multi-drug-resistant)? A study of more than six hundred people with TB in Cape Town, South Africa, indicated that 18 per cent were reinfected. Of these, 14 per cent were reinfected with a different strain. Similar results have been found in trials in China. It also appears that people who are reinfected are at a higher risk of developing active disease than someone who has never had the illness before. So exposure to TB doesn't appear to boost resistance to the disease.

Gillian Coates
Trefor, Anglesey, UK

Whorl class

? Do our fingerprints change? When I was a child I was taught that they never would. However, when I provided my biometrics for a visa application it was pointed out that my fingerprints did not match those that were taken five years earlier. Is this possible or is it more likely to be an error with the records or software?

Chisato Kobayashi
North Yorkshire, UK

Experts refer to the patterns that make up fingerprints as friction ridges. These begin to form in developing foetuses roughly halfway through pregnancy, and differ even between identical twins, and conditions in the uterus contribute to the pattern. With few exceptions, these patterns are permanent. Some people are born without fingerprints. As of 2011 members of five families are known to have adermatoglyphia, caused by improper expression of the SMARCAD1 protein. People with Naegeli–Franceschetti–Jadassohn syndrome and dermatopathia pigmentosa recticularis, which are both forms of the condition ectodermal dysplasia, also lack fingerprints.

Fingerprints can be temporarily erased by physical abrasion or medication, and because skin elasticity decreases with age, they can be difficult to record in older people. There are also criminals who try to evade detection by burning off their fingerprints or smoothing them out using glue and nail varnish. Notorious 1930s US gangster John Dillinger used acid in an attempt to erase his, and another, Robert Phillips, grafted skin from his chest onto his fingers to do the same – but was convicted using prints from other parts of his hands.

It has recently come to light that fingerprint analysis, once regarded as the gold standard in forensic science, is fallible. In his book *Finger Prints*, Francis Galton, cousin to Charles Darwin, calculated that the chance of a 'false positive' (two individuals having the same fingerprints) was about 1 in 64 billion. This may be true, but does not take into account that such analysis is not a perfect science, and relies on judgement, which is susceptible to cognitive bias. Fingerprint analysis is a twist on the spot-the-difference picture games we play as children, except here you are looking to match up as many as fifty 'landmarks' (for example where a ridge splits in two), often with a smudged or incomplete print from a crime scene. Assuming that you have not deliberately tried to erase your fingerprints, recent illness, your lifestyle, or occupation may have changed them. More likely one of the prints subjected to comparison was smudged or maybe there was a clerical error and one of the prints assigned to you belongs to someone else.

Mike Follows
Sutton Coldfield, West Midlands, UK

My hobbies include working on military vehicles, which involves cutting, welding, using knives, and sanding. My fingerprints suffer seriously. I have yet to lose a finger, but various scars show how close I have been. In a previous career, I worked on exploration oil rigs off the Egyptian coast. My job involved analysing cuttings brought up by the drilling fluid. Creating a microscopy slide meant fine grinding, invariably using a finger to rub the samples against emery paper. After a twenty-eight-day tour on the rig I had no fingerprints left, and it took twenty-eight days of leave to regrow them. So I'd say yes fingerprints can change. And although my laptop has a fingerprint reader, I stick to passwords.

Simon Mallett
Lenham, Kent, UK

Long in the tooth

? My dentist says that dental enamel is never replenished and we have to make do with the adult set of teeth throughout our grown-up lives. I can't believe that after some eighty years of chomping and grinding, my teeth are still the same shape, without apparent signs of wear. Is he right? If so, why am I generally still toothily intact?

John Everest
Pinner, Middlesex, UK

Despite the best efforts of your dentist, there is no regenerating lost enamel. By the time a tooth erupts through the gum, its enamel formation has ceased and it will have no further contact with the cells capable of repairing or remodelling it. So it is fortunate that the tooth's outer enamel layer is, in fact, the hardest tissue in the body. It is 96 per cent hydroxyapatite crystals by weight, the remainder being organic material and water.

Loss of this crystalline structure occurs in three main ways: attrition, the wear of enamel through tooth-to-tooth contact, such as grinding; abrasion, by external agents such as a hard-bristled toothbrush; and erosion, the demineralisation of the enamel crystals by acids.

Such acids can come from foods like fizzy drinks or citrus fruits, although stomach acid in those who experience gastric reflux can also be a culprit. Tooth decay results from a breach of the enamel due to acids produced by the bacteria in our mouth when we feed them sugars. As a result of all these processes, enamel will inevitably wear with age, although with due care and attention this can be kept to a minimum.

It is worth noting that fluoride applied to the surface of the teeth can strengthen enamel and make it more resistant to wear, but there is nothing available at the moment that can replenish lost enamel.

Stuart Yeaton
Oral and Maxillofacial Surgery Department, Barnet General Hospital, London, UK

Tooth enamel is never replenished because the ameloblast cells that form the enamel are on the surface of the developing tooth. These cells are destroyed when the tooth erupts into the mouth.

Tooth enamel (calcium hydroxyapatite) is the hardest material in the body. By contrast, the Western diet is generally very soft and rarely contains any abrasive materials. Consequently very little attrition occurs during whatever minimal chewing is needed. What's more, under normal conditions, teeth only fully come into contact with those in the opposing jaw when we swallow. During chewing, food in between the teeth reduces tooth-to-tooth contact.

Between meals, when the jaw is at rest, there is normally a gap of 2 to 3 millimetres between the teeth – the freeway space – so no attrition takes place. Over the course of our lives, therefore, the teeth are not in contact for much of the time and so wear and tear is minimal.

However, under psychological stress, tooth grinding and clenching can lead to excessive wear. One common source of extreme wear is bruxism, the unconscious grinding of the teeth while we are asleep.

Philip Taylor, retired dental surgeon
Kinloss, Moray, UK

Head space

? My thoughts feel as though they are in my head.
Is there a physical basis for this or is it just that I
know that's where my brain is?

Julian Richards
London, UK

Your head is the focus for receiving information from your
environment, mainly through sight and sound. Thought is
secondary to sensory perception, but both have to combine
quickly when you are threatened, and evolution has sited the
organs for sensing information and processing it close
together to ensure speedy response to dangers such as sabre-
toothed tigers and muggers.

Researchers at the Karolinska Institute in Stockholm,
Sweden, investigated out-of-body illusions using a brain
scanner. They found that the brain's posterior cingulate cortex
combines the feelings of where the self is located with that
of body ownership. This is why you feel as if your thoughts
are inside your head. Sensory deprivation can cause such
feelings to be lost. When people lose sensory awareness while
they are still conscious they become disorientated and may
experience out-of-body sensations feeling as though their
thoughts are no longer inside their head. This can occur
during certain types of torture or can be self-induced through
meditation or drug use.

David Muir
Science Department, Portobello High School, Edinburgh, UK

Rumble of doubt

 How are internal body noises such as stomach
rumbles produced, and why can we hear them?

Jeremy Smith
Plymouth, Devon, UK

There is always gas in the human gut, mostly hydrogen,
nitrogen, carbon dioxide and methane. This gurgles about in
a semi-liquid substrate. Vibrations in our gut generate audible
sounds just like the skin of a drum does. Central-heating
systems also produce gurgling noises, which are caused by
trapped gases. The noises go away after 'bleeding' the radi-
ators to release them. We bleed our stomach and bowels by
burping and farting.

Hugh Hunt
Cambridge, UK

Resist and multiply

 We hear a lot about bugs, flies and fungi gaining
resistance to pesticides and other chemical
treatments, making such solutions futile. If this is the
case, shouldn't humans gain resistance to threats such as
harmful chemicals and bacteria?

Nimesh Nambiar
Dubai, United Arab Emirates

Humans accumulate genetic changes all the time, as has
become obvious through whole genome sequencing studies.
These can range from as little as a single base substitution – a

change to one 'letter' of the 3 billion pairs in your genome – to alterations to whole chunks of DNA. But because the human genome is so big and complex, many of these changes have no discernible biological consequences.

Known adaptations to threats include certain populations that have retained the genes for blood diseases such as thalassaemia and ovalocytosis, because these confer some resistance to malaria, a more serious threat. However, if an advantageous mutation occurs, it can take a long time to spread through the population, because humans live for many years and produce relatively few offspring. This generation time has a key role in determining the rate at which a genetic change gets established in a population, but genome size is also crucial.

As organisms become less complex down the evolutionary tree, their genomes become smaller. Our genetic material encodes approximately 30,000 genes known so far. Insects, such as the mosquito that transmits malaria or yellow fever, have just over 12,000 genes. Parasites, such as malaria, are unicellular organisms that have around 5,300 genes. Bacteria are even smaller.

The relevance of this is that the smaller and simpler the genome, the more likely it is that changes will have biological consequences. This means that random changes can be more significant and make a survival difference, but for the change to have an impact in the population, it has to be propagated through subsequent generations.

The propagation of genetic changes depends very much on the lifespan of the individuals and on how many offspring they can transmit this change to. Humans commonly have no more than a few offspring in their lifetime, whereas mosquitos can lay hundreds of eggs. Unicellular parasites, such as malaria, have a life cycle of forty-eight hours in the blood. During this time they can proliferate exponentially, so any genetic change that gives a survival advantage spreads very quickly.

This is why parasites such as malaria are notorious for developing resistance to drugs, and why mosquitos become less sensitive to insecticides, but we humans are slow to adapt to our own biological threats.

Alena Pance
The Wellcome Trust Sanger Institute, Cambridgeshire, UK

Fishy hell

> **?** I don't like eating fish and I find their odour disagreeable. My wife says it's all in my head, but I think that many people have my aversion. Is there a genetic or evolutionary component to it, is it based on an unpleasant childhood experience, or do I have to concede to my wife's theory?

Al Morin
London, UK

It's common not to be able to pinpoint the origin of someone's distastes. Some may be partly genetic. For example, people to whom the chemical phenylthiocarbamide (PTC) tastes bitter are genetically unable to taste the bitterness of the fruit of the Chinese laurel (*Antidesma bunius*), and vice versa.

Most often, however, one acquires preferences for tastes and smells that one grows up with, and distastes in reaction to bad associations, particularly in early childhood. Having loved the taste of kidneys at first, I vomited up a kidney meal at the age of six, when I had scarlet fever. It took me decades to like them again.

As for fish, Richard Feynman once related how he used to hate fish as a child, but liked it as an adult when he tried some in Japan. Back home, he found that he still hated fish. It turned out what he hated was the taste of fish that was not quite as fresh as available in Japan.

Jon Richfield
Somerset West, South Africa

In my case at least, fish aversion seems to have been the result of an unpleasant childhood experience. This was only revealed to me when I revisited my childhood home after many years, and talked to one of our neighbours. 'I remember you,' she recounted. 'Your mother often left you in the care of your older brothers, who bought fish and chips for lunch. The three of you then sat in the gutter, where they ate all the chips and forced you to eat the fish.'

Subsequently I never ate fish and became almost nauseous from the smell, which I now know to be associated with unfresh fish. Eventually I was so embarrassed at asking for an alternative dish at dinner parties that I steeled myself for a fish course, which turned out to neither smell nor taste as I was sure it would.

Since then fish has become a real delicacy for me. That includes oysters (once an absolute anathema) as well as raw fish, which, curiously, has little flavour.

So I have to agree with your correspondent's wife that his aversion is probably the result of nurture rather than nature.

Don Ross
Vaucluse, New South Wales, Australia

Bloodthirst

? Do humans have an innate desire to eat meat, or is it cultural? If our culture had all references to eating meat removed, would people still desire it?

Richard Brown
Geneva, Switzerland

It is less an innate desire for meat in particular than for concentrated nutrition such as fats and proteins. Even in non-meat foods, we favour concentrated items such as nuts, fruits, grains and tubers. Think of how appetising chocolate or halva is, compared with lettuce or grass.

Herbivores generally must eat large quantities of low-grade plant foods, discarding most of the fibre and excess materials they contain. They also discard the toxins; plants are generally full of harmful chemicals that herbivores must tolerate, discard or destroy, while concentrating the vitamins, proteins, fats and digestible carbohydrates.

The resultant purified herbivore flesh suits carnivores that would die if they tried eating too much of the wrong plant matter; the list of wholesome plant feed that could kill your dog or cat is shocking. Many herbivores are partial to a bit of meat if they can get it, and omnivores generally will work harder for animal food than plant food.

So it is with us; most of us relish and thrive on some of the concentrated fat, protein, vitamins and essential fatty acids in meat.

Jon Richfield
Somerset West, South Africa

Light snack

? I am a vegan so my body gets a lot of its vitamin D when I am exposed to the sun. Must the sunshine be absorbed 'directly' or can I get the benefit when it arrives through glass windows? How much do I need per day?

Colin Brown
Edinburgh UK

Vegan or omnivore, most of us depend on sunlight for meeting our vitamin D needs. And it is not strictly a vitamin, because adequate sunlight can supply your entire requirement without a dietary input. Estimates of daily requirements vary, but UK guidelines suggest 10 micrograms for adults who cannot get enough sunlight, such as those with cultural reasons for covering their body and face, and for elderly people confined indoors. Similar levels are recommended in the US.

Good sources of dietary vitamin D are oily fish, eggs and liver, but these are not consumed frequently enough to ensure high intakes. Despite fortification of foods, including breakfast cereals and spreadable fats, the average dietary intake in the UK is poor – somewhere between 2 and 3.5 micrograms daily. A young fair-skinned adult should be able to synthesise enough vitamin D by exposing their face to the sun for about twenty to thirty minutes, but elderly people and those with dark skin will need more time. Older adults produce about 25 per cent of the levels young adults do for the same length of exposure.

The amount of exposed skin is also a variable, so naturists should find it easier to meet their daily allowance. Sunlight exposure is not, however, as straightforward as it seems, because only the narrow band of ultraviolet B (UVB) light (between 290 and 320 nanometres) is suitable for synthesis of vitamin D in the skin. The further you live from the equator, the harder it is to get an effective dose of sunlight, especially in winter. Winter levels of vitamin D in the blood are half those of the summer. If your shadow on the ground is longer than you are tall, then you cannot synthesise vitamin D (this is known as the 'shadow rule'). In Scotland for example, early morning sun in December is useless and there is no significant synthesis of vitamin D between October and March. In summer it would take someone in London around twice as long to make the same amount of vitamin D as a person sunning themselves in Barcelona, Spain.

But to answer the question, you need to get outside in the summer, because little UVB will pass through glass unless you invest in sapphire glass. It is not possible to overdose on vitamin D from sunlight, because erythema or reddening of the skin coincides with its breakdown, so there's no need to go to extremes in the summer. However, in rare cases, too much dietary vitamin D can be harmful, so supplements should be taken with care. Astronauts, submariners and polar workers need supplements for obvious reasons.

Brian Ratcliffe, Emeritus Professor of Nutrition
Robert Gordon University, Aberdeen, UK

Nutty question

? Why is the scrotum so wrinkly?

Neil Smith, via email

The scrotum plays a valuable role in thermoregulation of the testicles, as sperm production is reduced at core body temperatures and above. Those of us with this piece of anatomy will have noticed that on a very cold day, or when emerging from a cold shower, the scrotum (and the rest) will be much smaller. The converse happens on a warm day or on emerging from a hot bath. The scrotum is quite smooth when relaxed and only wrinkles as it is pulled tight to the body.

This is because muscles within the scrotal wall contract to bring the testicles nearer the abdomen in cold conditions, and relax to keep them cool when it is hot. There is also a complex system of vasculature in the spermatic cord (the pampiniform plexus), where warm blood leaving the abdomen is cooled by a heat exchanger system before entering the testicles.

When farmers first started to rear merino sheep in Australia, they found that the rams were sterile until sheared, so their testicles are shorn on a regular basis today. This breed of sheep originated in southern Portugal, where summer temperatures are not as high. Consequently, the rams have unsuitably woolly scrotums for southern Australian conditions.

Bob Butler, retired veterinarian
Llangoed, Anglesey, UK

Light as air

? While on the scales this morning I wondered, would passing gas affect the weight of the human body at sea level and, if so, in which direction?

Chris Gilfillan
Surrey Hills, Victoria, Australia

Flatulence is mostly methane, hence the schoolboy trick of putting a match to it. Methane (molecular weight 16) is lighter than air, which is approximately 80 per cent nitrogen (molecular weight 28) and 20 per cent oxygen (molecular weight 32). So you will gain weight on the scales when you pass wind. The odoriferous components are all denser (the lightest would be hydrogen sulphide, molecular weight 34), but are not likely to be present in sufficient quantity to make much difference.

Burping after a fizzy drink will have the opposite effect, because you are expelling carbon dioxide (molecular weight 44), which is denser than air. So your weight on the scales will decrease.

Guy Cox
School of Medical Sciences, University of Sydney, New South Wales, Australia

It's good to talk

? What were the first words, and who spoke them?

It's a fair guess that there was once an original mother tongue – the ancestor to all living and dead human languages. The evidence for this is that all human languages, unlike other forms of animal communication, string together words into sentences that have subjects, verbs and objects ('I kicked the ball'), and anyone can learn any language.

Comparative linguists search for sounds that come up again and again in languages from all over the world. They argue that if any relics of a mother tongue still exist today, they will be in those sounds. Merritt Ruhlen at Stanford University in California, for example, argues that sounds like *tok*, *tik*, *dik*, and *tak* are repeatedly used in different languages to signify a toe, a digit or the number one. Although studies by Ruhlen and others are contentious, the list of words they say are globally shared because they sound almost the same also includes *who, what, two* and *water*.

Another approach is to look at words that change very slowly over long periods of time. My own team has used such statistical studies to show that words for the numbers 1 to 5 are some of the slowest evolving. Also on this list are words involved in social communication, like *who, what, where, why, when, I, you, she, he* and *it*. This list fits with the expectation that language evolved because of its social role. It also has some overlap with Ruhlen's list.

More broadly, we can say with some confidence that the first words probably fitted into just a few categories. The first ones may have been simple names, like those used by some of our primate relatives. Vervet monkeys give distinct alarm calls for leopards, martial eagles and pythons, and young vervets must learn these. In humans, *mama* is a strong candidate for a very early noun, given how naturally the sound appears in babbling and how dependent babies are on their mothers. The sound '*m*' is also present in nearly all the world's languages.

Imperatives like *look* or *listen* are also likely to have appeared early on, perhaps alongside verbs like *stab* or *trade* that would have helped coordinate hunting or exchanges. Even this simple lexicon allows sentences like 'look, wildebeest' or 'trade arrows'. Finally, simple social words like *you, me* and *I, yes* and *no*, were probably part of our early vocab. Amusingly, a recent study suggested that *huh* is universal, prompting headlines that it was among the first human words. Perhaps it was the second.

Mark Pagel
School of Biological Sciences, University of Reading, UK

Deposit account

? Our family of five eats essentially the same meals and lives in the same environment yet our faeces have different and identifiable smells depending on the individual. This must be to do with whatever lives in our intestines. But what is it?

Gaynor (aged twelve), Adam (aged nine) and Serena (aged seven) Randall
Birmingham, UK

The characteristic odour of faeces is almost entirely due to the presence of minute quantities of aromatic nitrogen compounds such as pyridines, pyrazines and indoles. These are present in varying proportions in the faeces of different individuals and constitute a sort of olfactory fingerprint. As writer G. K. Chesterton noted in *The Song of Quoodle*, the human nose is notoriously insensitive compared with most mammals.

However, it is extremely sensitive to these particular compounds, enabling us to distinguish our own faeces containing microflora to which our bodies are accustomed, from those of other people or animals which may contain dangerous pathogens. This is why we find the smell of our own stools and flatulence relatively inoffensive while those of other people are nauseating. Interestingly, the perfumery industry takes advantage of this sensitivity using trace amounts of these compounds to add piquancy to its products.

Roger Savidge
Shoreham by Sea, West Sussex, UK

Picked clean

? Although it seems to be frowned on in polite company, is it in fact necessary to pick your nose to keep it clear and functional? Nobody likes their nose to be dirty or uncomfortable, so why is nose-picking such a taboo area in many cultures?

Gavin Bate
Carshalton, London, UK

I don't imagine it is necessary to keep nasal passages clear at all times. The annoyance experienced by people with a partially blocked passage suggests a deeper story. Somewhere in our past, there may have been nasal parasites (adult, larvae or eggs) we would have been smart to eliminate before they travelled further into the sinuses. A clean nose was a safe one.

There's a stronger case for the taboo of nose-picking. It fits with a more general taboo against bringing hidden bodily tissue or products into sight. We swish saliva around our mouths all the time, yet react with disgust when it emerges as spit or drool. And it's the same for earwax or dandruff.

'Wrong things in the wrong place' can cause contagion and they are full of microbes, even if not necessarily pathogenic ones. You don't have to be in danger of contracting Ebola to be wisely wary of bodily fluids and nasal contents.

Louise Fabiani, via email

I know that face

? I've seen several archaeological shows in which skulls are built up to create 'real-life faces'. How valid is this? Has anyone tested the procedure by giving people a complete, contemporary skull when a picture exists of the person, and seeing how it compares to the recreated image?

Ian Wardrop
Houghton, Adelaide, South Australia

Having developed the Manchester Technique of Forensic Facial Reconstruction at Manchester University in the UK between 1973 and 1980, I can confirm that it is perfectly valid. The face will automatically grow outwards from the cast of a skull if you develop the muscles and facial structures using clay or wax, guided by average soft tissue measurements.

The technique was put to the test when I began assisting with police investigations into the identity of human remains. Using this reconstruction method, approximately 60 per cent of the individuals were recognised. In these instances, a reconstruction could be checked with an image of the subject. Once a name was provided, more conventional techniques could be applied to confirm the identity.

It soon became clear that although such studies would never result in an exact portrait, they would routinely result in a face similar enough to the subject's that the correct identity could be ascertained. By today's standards this approach may seem rather simplistic, but one must remember this was before DNA testing became available.

Over the course of my career I have undertaken a number of studies, some on casts of the skulls of living people created using scanning technology. A similarity with the subject can always be observed. Although this work was regarded with some suspicion when it first started back in the 1970s, today it can be studied at degree level and is one of many techniques available for both forensic and archaeological investigations.

Richard Neave
Rye, East Sussex, UK

Uncommon ancestor

? Like other white Europeans I carry about 2.5 per cent Neanderthal genes, more than indigenous Africans. If we all walked out of Africa about 100,000 years ago, what is the explanation for this?

A. M. Bryant
La Réortière, Le Busseau, France

We didn't simply walk out of Africa at any particular time. We originated there, and our ancestors sallied forth everywhere, but many genetically distinct populations took various routes over hundreds of thousands of years and their paths repeatedly diverged and converged.

In palaeontological terms, any event that lasts only for millennia is pretty abrupt. Our prehistory back to the estimated origin of agriculture probably spans less than 15,000 years. We are a recent, transient rabble – yet to us, family histories of 1,000 years seem ancient.

The Neanderthals were early colonists in Europe, sufficiently distinct from other populations of the genus Homo that geneticists can tell them apart. This factor has given us a good idea of when and to what extent Neanderthals merged with their successors. Perhaps hundreds of thousands of years after the Neanderthals, invaders from Africa introduced new genetic material: those that settled in Europe interbred with Neanderthals, so the proportion of such DNA in their descendants is greater than in other populations.

Jon Richfield
Somerset West, South Africa

Crying foul

? Babies in TV or films are frequently depicted crying. How are they made to do this? Does the director wait for the child to cry normally, or is there some scientific method that ensures the youngster doesn't feel pain?

Ken Lee
Hervey Bay, Queensland, Australia

I'm a practising midwife and the midwifery adviser for all five series of the BBC's *Call the Midwife*, so I have a lot of experience in handling babies on film sets. Film sets are not easy places for tiny people, and if they cry it's not usually because that is what the director requires, it is just what happens.

The regulations around the presence of babies on a film set are understandably very restrictive. Each scene takes a long time to set up. The baby is invariably only required for a few minutes at a time, which means that baby, parent and their chaperone are waiting in a room away from the set until they are called. While every effort is made to ensure the set is baby-friendly – calm, warm and quiet – this can't always be achieved. For example, costumes can cause problems because most babies don't like being disturbed to be dressed or undressed. A costume change is done as late as feasible, then the baby is kept with their parent until the last possible moment. This helps to keep both the parent and baby content.

The film set is usually an unfamiliar environment for the parent. Although they are reassured and can watch the baby on a monitor, stress levels are still quite high. The actor will have been briefed, rehearsed and where possible will have met the baby before filming the scene, but sometimes this will be the first time they have even held an infant. Babies quickly pick up adult feelings and emotions, so an anxious actor usually ends up holding a crying baby. Babies are not made to cry on cue, so any cry that is caught on camera is serendipitous.

Terri Coates
Salisbury, UK

Selected selections

? The only genetic changes in humans we ever hear about are those producing diseases such as cystic fibrosis. Has anyone identified any genetic changes within recent generations that make individuals possessing them 'more fit' to thrive in today's environment? Would we even know these changes if we saw them? And would we consider them normal for healthy humans?

Colin Bamforth
Altrincham, Cheshire, UK

The example that is most often cited of a positive genetic change within a recent generation is that of sickle-cell anaemia, which appears to give an advantage to people who have it in places where malaria is endemic. While we wouldn't necessarily consider having sickle-cell anaemia as a healthy trait, carrying a copy of the 'faulty' gene confers a level of resistance to malaria, and is therefore selected for in populations exposed to malaria parasites. This is generally taught in most high-school biology courses as an example of human natural selection, as well as the idea that a trait is neither positive nor negative intrinsically: rather its value is dependent on the environment.

However, another example of a positive trait that is necessary for survival in today's environment is our adjustment to being able to cope with the high level of sodium in our diet. In populations that have a low daily salt intake, such as some indigenous Amazonians, people can retain salt effectively; they lose less sodium in their sweat and urine compared with average citizens of the US. If they are subsequently exposed to a high-sodium diet, however, this ability to retain salt works against them and many may suffer from hypertensive disorders, and die at a very young age from heart disease.

Mark Bilger
Livonia, Michigan, US

Green genome

? If we could somehow splice the recipe that plants use for making chlorophyll into our genes, could we then satisfy some of our energy needs by photosynthesising?

Sophie Holroyd
Bromyard, Herefordshire, UK

There are three main hurdles. First you would need to make the chlorophyll, then deploy it to harness energy from light for synthesising ATP and NADPH, and then use these metabolites to convert carbon dioxide into sugar. Surprisingly, we already have many of the genes needed for the first and last stages.

Plants use sixteen metabolic reactions to create chlorophyll, nine of which are shared with the pathway we use for making haem, a constituent of red blood cells. So inserting genes for seven extra steps could in principle allow us to make chlorophyll. Unfortunately, the chlorophyll would be toxic to humans, especially in sunlight, unless we also copied plants in making special proteins that envelop the chlorophyll molecules.

The last stage, known as the Calvin–Benson cycle, requires genes for eleven enzymes, nine of which we already have. The two missing enzymes are RuBisCO and phosphoribulokinase.

The middle stage is the trickiest because, although we already have the enzymes needed to make ATP and NADPH, we don't have a way of powering them by light energy. Plants have ingenious proteins and lipids, structurally organised within chloroplast membranes to allow this.

Before you try to engineer these botanical feats into your own chromosomes, be warned that you would be able to photosynthesise only minute amounts of carbohydrate compared with what you get from, say, a slice of bread. The problem is that your body would need a huge surface area relative to its volume in order to absorb useful amounts of light. Obviously, the chlorophyll would need to be in your skin; your liver, lungs and brain are not well placed to absorb light. I weigh 85 kilograms and my skin provides a surface area of less than 2 square metres. In contrast, an 85-kilogram plant typically presents an area of over 200 square metres to the sun, thanks to its thin, expansive leaves. With an area to volume ratio like that, we would have great difficulty moving around – in fact, we'd be vegetative.

Stephen C. Fry, Professor of Plant Biochemistry
School of Biological Sciences, University of Edinburgh, UK

5 Life on Earth

Furry berry

[?] My six-year-old daughter Victoire wants to know why raspberries are covered with little hairs, unlike other berries.

M. Blanc
Lyon, France

The hairs on raspberries are the remains of the female parts of the raspberry flower, which have not fallen away. In the flower, the female hair-like styles are collected in the centre with the male anthers arranged around the edge. Each style, topped by a stigma, is connected to one ovary, forming a pistil.

After pollination, the petals, anthers and other parts of the flower wither away, and each ovary swells to produce a segment of the final fruit. Each of these segments is, botanically speaking, a drupelet, and a raspberry is a drupe (a collection of these segments) rather than a berry.

The plants that I grow in my garden often have raspberries at all stages of development, from flower bud to ripe fruit, so the full cycle can be seen clearly throughout the same plant.

In fact, raspberries are very easy to grow, and very productive, but you do need to take steps to contain them.

Dylan Brewis
London, UK

Sharp taste

? Why do plants with berries often have thorns? Aren't such plants sending mixed messages? Why would they want to repel animals that would eat their berries and spread their seeds?

Marinus Lutz
Vancouver, Canada

The messages aren't mixed: the plants are saying 'eat my fruit, not my leaves' . Most animals that benefit the plant by eating fruit and dispersing seeds are dexterous enough not to hurt themselves. A bigger, clumsier herbivore might not bother separating fruit from foliage, but if the latter came with a mouthful of thorns, the animal might think again.

A related issue is why some tasty fruits have highly distasteful skins – oranges, bananas and mangoes are good examples. To enjoy these, an animal has to be big and powerful enough to peel the fruit, and thus probably capable of eating the peeled fruit whole. This means swallowing the contained seeds and depositing them many miles away, rather than just nibbling the juicy bits and leaving the seeds unhelpfully close to the parent plant.

This is similar to the reason that unripe fruits are small, inconspicuously coloured, sour and rich in unpleasantly astringent tannins. The situation changes when the seeds are fully developed and could benefit from being dispersed.

Stephen C. Fry, Professor of Plant Biochemistry
School of Biological Sciences, University of Edinburgh, UK

I have found that larger animals, such as horses, prefer to eat the stems of plants that produce berries, thus damaging the structure of the plant. Thorns are a helpful deterrent. Creatures that prefer the berries and do not damage the rest of the plant, such as birds and insects, are able to bypass the thorns to eat the berries and help spread the seeds.

Cordelia Moore
Ashurst Wood, West Sussex, UK

Tree time

? It is generally accepted that the increase in day length in the spring is what prompts deciduous trees to begin growing their leaves. But how does a tree know that the days are getting longer?

Ray Sheldon
Bridgetown, Nova Scotia, Canada

Back in 1920, botanists Wightman Wells Garner and Harry Ardell Allard thought the length of daylight was the key to photoperiodism. So by the time it was discovered that the hours of unbroken darkness were responsible, plants had already been labelled as 'long-day' 'short-day' or 'day-neutral'. Chrysanthemums are short-day plants, and will only flower if there are enough hours of darkness. Growers of these plants used to keep their greenhouse lights on all night to delay flowering until it was discovered that just a brief burst of light would suffice.

This is down to the photosensitive phytochrome pigment. It exists in two states, and when exposed to sunlight the amount of pigment in either state is about the same. As night falls, more of the pigment switches to the active state.

But a flash of light will redress the balance, and this is enough to postpone flowering in short-day plants or induce it in long-day plants. Even seeds are armed with this pigment, which allows them to use the direction of sunlight to sense which way is up (this ensures shoots and roots head off in the right direction). They can 'calculate' how deeply they are buried from the light intensity, and can even detect the presence of overhanging foliage and postpone germination.

Mike Follows
Sutton Coldfield, West Midlands, UK

Everyone knows that plants need light to grow and that most plants grow in different ways at different times of the year, particularly in temperate latitudes. This effect, called photoperiodism, enables temperate plants to control when they grow new leaves and when to drop them in the autumn. However, it isn't the length of the day which is important, but the uninterrupted dark period – the length of night. During the period of darkness the plant produces a photosensitive protein called phytochrome, the concentration of which controls the onset of dormancy in autumn and bud-burst in spring. Even a small amount of electric light during the hours of darkness can upset this, and it is noticeable when a tree encroaches on a street light. In autumn, when the tree is starting to become dormant, the leaves closest to the light are often the last to fall. However, trees and plants are unaffected by light from the moon.

John Crofts
Sherwood, Nottinghamshire, UK

Gimme the plane truth

? I have often been told that plane trees in London can cope with the high levels of pollution in the city because they draw it into their bark, which they then shed. Is there any truth in this?

Mike Griffiths
Croydon, Surrey, UK

Plane trees are particularly numerous in London because they are able to remain healthy despite high levels of pollution. It was for this reason that large numbers of them were planted during the Industrial Revolution, when pollution in the city was particularly high. Many survive to this day.

All trees absorb gases through their trunks. Their bark is full of tiny pores called lenticels. These allow air to reach the wood inside where gases such as oxygen are vital for processes like respiration. In cities, these pores often become blocked with pollutants from factories and vehicles and these become visible as black dots on the bark. What sets plane trees (and some other species like birch) apart is the rapid rate at which fresh bark grows beneath the outer layer. This means that trees can quickly shed their outer layer of bark which is laden with pollutants. It is this shedding process that gives the trunks of plane trees their mottled appearance.

Lewis O'Shaughnessy
Salisbury, Wiltshire, UK

She smells sea smells

[?] Does the sea smell of fish or do fish smell of the sea? What is responsible for the characteristic seaside smell and was it present in the chemistry of the oceans before the first fish evolved?

Tom Curtis
London, UK

A newly caught fish rinsed in fresh water does not have a noticeable aroma. The smell of the seaside is caused by a cocktail of chemicals, principally dimethyl sulphide (DMS), which people can smell at concentrations as low as 0.02 parts per million. Phytoplankton are single-celled organisms found in the sea and use energy from sunlight to make dimethylsulphoniopropionate (DMSP). This chemical is consumed by marine microbes and some is converted into DMS. Because phytoplankton have an earlier origin than fish, and are at the bottom of the food chain, the characteristic seaside smell was around long before fish swam onto the scene.

DMS molecules act as condensation nuclei for clouds, and James Lovelock was one of those who proposed that it might form part of a negative feedback loop – helping to regulate our climate according to his Gaia hypothesis. As more sunlight reaches Earth, it increases the surface temperature. But it also increases the phytoplankton population so the production of DMS increases and more clouds are seeded. These clouds reflect sunlight, reducing both the surface temperature and the phytoplankton population.

Increased cloudiness is also associated with higher wind speed over the oceans. This mixes surface waters, bringing up nutrients for phytoplankton in readiness for a new burst of sunshine. So when you next stroll along the beach and breathe in the sea air, you might want to reflect on the ocean biochemistry that helps to keep Earth's temperature suitable for life.

Mike Follows
Sutton Coldfield, West Midlands, UK

Algal bloom

? If Earth were to be suddenly devoid of all life but otherwise unchanged, and I were to drop a single photosynthetic bacterium or alga in the ocean, how long would it take to colonise all the seas? And how long to colonise the land?

Alain Williams
London, UK

Algae, which is Latin for 'seaweed', is a diverse group of species ranging from single-celled organisms to the 50-metre-long giant kelp *Macrocystis pyrifera*. Like plants, algae are autotrophs and make their own food, harnessing sunlight to synthesise sugars from carbon dioxide and producing oxygen as a waste product. Imagine starting with an individual *Pediastrum boryanum* which is about 10 micrometres across. If it only reproduced asexually (by dividing in two) it would take eighty-five generations to cover all the oceans. If each generation reached sexual maturity in twenty-four hours, this process would take about three months.

Algae are naturally aquatic, so the colony would need to wait for the development of another species with which it could develop a symbiotic relationship before populating the land, as has fungi with lichen. In reality, individual blooms appear to form quickly because algae are fairly ubiquitous but don't become noticeable until the population density is sufficient to make the water semi-transparent, with the population increasing a thousand-fold in just ten days. Of course, the population density that defines a bloom depends on the species of algae, which also lead to different colour blooms. Some blooms like the so-called red tide are toxic and a threat to the local ecosystem.

Even in the absence of microscopic herbivores nibbling at its progress, the geographical reach of blooming algae would be curtailed because its requirements for sunlight, warmth and nutrients aren't present everywhere at any given time. Indeed, algal blooms are often associated with pollution events, especially where nitrogen and phosphorus from fertiliser runs off into the ocean. A lack of nutrients at latitudes between 30° north and south of the equator is why tropical waters are often referred to as oceanic 'deserts'. This results in the unusually clear blue water that attracts scuba divers.

Mike Follows
Sutton Coldfield, West Midlands, UK

First let's make some assumptions for a simple model, some of which ease the computational task. We'll work with powers of 2, which means we only multiply or divide by 2. Thus we'll round up Earth's surface area from about 510-million-square kilometres to 512 million. Until a biologist provides a different figure let our alga be 1 square micrometre in size. Finally, a common school question has lily pads doubling in area every twenty-four hours, so we'll use this as well. So how many days does it take to go from 1 square micrometre to 512 million square kilometres, if our algae doubles in area every day?

In ninety days it would easily cover the surface area of our model Earth. That would include the surface area in many caves and all other land features. Using a rough figure of 75 per cent for the area of Earth covered by sea our alga would still need the same number of days to colonise the oceans as the entire planet as day eighty-nine represents 50 per cent planet cover. However, like any model, this needs to be taken with a pinch of salt. A few things that could easily happen to render the model invalid include the possibilities that the alga may only grow in salt water so cannot cover land or fresh water; that death from starvation may slow growth; or that other outside factors such as inappropriate climate in some locations may also slow growth. Finally, if 1 square micrometre seems big for an alga, changing its size to 1 square nanometre adds just ten days to the model.

David Morton
Geeveston, Tasmania, Australia

Pond skimmer

I have a garden pond that suffers from 'pea-soup' algae. A thunderstorm recently flooded the nearby road, overflowing into my garden and the pond. The next morning the algae had gone. What was in the floodwater that managed to clear the algae?

Colin Carter
Marlborough, Wiltshire, UK

'Pea-soup' algae are generally single-celled species, plus others that clump in twos and fours. Populations explode when sunshine and nutrients combine in warm conditions – *Daphnia* and similar organisms that graze on algae are barely able to thin the ranks. The resultant pea soup may look repellent, but is often a healthy community that is enthralling to view under a microscope. To clear the water, key nutrients such as phosphates must be reduced.

Floodwater might wash out a pond or cause the algae to die through pollution. Dead cells will sink or form a scum, leaving the rest as clear water. Possible toxic pollutants include roadside herbicides, salt or tarry phenolics from asphalt on recently surfaced roads. Furthermore, heavy traffic sprinkles roads with exhaust dust and rubber powder. Tyre rubber contains toxic organic compounds and metals such as zinc, with the dust having a huge surface-to-volume ratio that permits rapid release of poisons into the water. In fact, it is a common (although ill-advised) practice to put old tyres into ponds to clear them of unwelcome organisms. However, it actually increases the nutrient supply and invites cyano-bacterial blooms that are far less welcome than any pea soup.

Jon Richfield
Somerset West, South Africa

Algae do well in a neutral to slightly alkaline environment, at a pH of about 7 or 8. At this level they take up nutrients such as phosphate and nitrate easily.

All rain is acidic. Carbon dioxide makes up about 0.04 per cent of air and it dissolves in clouds and falling rain to form carbonic acid, giving natural unpolluted rain a pH of about 5.5. Air-polluting gases such as the oxides of sulphur and nitrogen further reduce the pH of rain.

Normal amounts of rainfall on the surface of a pond would decrease its pH, but it is buffered by carbonates and bicarbonates – naturally occurring alkaline chemicals in water. However, a deluge of rainwater would have defeated the pond's buffering capacity, lowering its pH.

An acidic pH can damage algal cell walls, thus causing the

overnight algal demise in this case. Because pond plants use up the excess carbon dioxide during photosynthesis, the pond's pH will start to return to normal. But pond owners should not be tempted to use acid to get rid of algal blooms because it could have disastrous consequences for pond flora and fauna.

David Muir
Science Department, Portobello High School, Edinburgh, UK

As photosynthetic organisms, algae will tend to cluster close to the surface of the pond to capture maximum light (just as trees in a dense woodland have most of their leaves at the top). So, although the water may appear as 'pea soup', it is probably much clearer towards the bottom.

The storm water pouring into the pond causes it to overflow. And the overflow comes from the top layers of water. It is also possible the storm water, having fallen from a height and being colder than the pond water, sank rapidly, enhancing the effect. So perhaps it is nothing in the storm water itself. Within a week, I'll bet it turned just as pea-soupy as it was before.

Alistair Scott
Gland, Switzerland

Slick thinking

Why are dispersants used to treat oil spills? Wouldn't a coagulant simplify containment and clean up?

Howard Bobry
Nehalem, Oregon, US

Coagulants are sometimes used to treat turbidity, or cloudiness in water. This causes the particles responsible to coalesce, or flocculate, making them easier to remove by precipitation or filtration. When oil is in water, the two form a colloid – a solution in which one substance of microscopically dispersed insoluble particles is suspended throughout another. Colloidal particles do not usually flocculate on their own, because they often carry an electric charge, thus repelling each other.

Coagulants carry an opposite charge to neutralise the colloid, allowing flocculation. The hydrocarbon molecules that constitute oil spills are non-polar, carrying no electric charge, so do not mix with electrically polarised water molecules or coagulants. Dispersants such as detergent have a split personality. Their molecules have a long non-polar tail and a short electrically charged head. The tails dissolve in oil to form microscopic droplets called micelles, from which heads stick out into the water, allowing mixing and thus easier dispersal. The micelles then disperse to hopefully be biodegraded.

However, oil micelles can cause damage. You need look no further than the economic and environmental problems caused by the Deepwater Horizon disaster of 2010 in the Gulf of ·Mexico. About 7 million litres of dispersant were used to combat this spill. Sadly, a dispersant can be more toxic than the oil it dissipates – and worse, can act in synergy with the oil, causing toxicity many times greater than the sum of the individual toxicities of the oil and dispersant.

David Muir
Science Department, Portobello High School, Edinburgh, UK

Poultry question

? Are we making the threat of avian flu worse by killing infected birds? Would it make more sense to isolate them and let them develop immunity naturally? Or is bird flu so serious that they would all die anyway?

Richard Hind
York, UK

The idea is not to save the lives of the birds. It is to prevent the spread and evolution of the virus. Helping a few birds develop immunity would not stop the virus from evolving or spreading to other birds or to us. Many birds would be likely to survive anyway. But since the idea is to kill the virus we have to try to kill all the affected birds.

Eric Kvaalen
Les Essarts-le-Roi, France

Your previous correspondent's reply to this, although correct as far as it goes, does not fully address the original question. The reason for mass culling is not so much to eliminate the virus as to minimise the risk of it spreading to humans. Breeding resistant poultry would completely eliminate the virus and in a more effective way than culling flocks. However, flu viruses mutate very quickly, so a breeding programme might be too slow. We take a different approach to plant crops. When wheat is infected by rust fungus we do not destroy the whole crop, but replant the following year with a resistant variety.

Guy Cox
St Albans, New South Wales, Australia

Cold calling

[?] Vaccination against flu viruses is well established and reasonably successful, especially in high-risk groups such as the very young and elderly. Colds are also caused by viruses, and although they are less virulent they occur more frequently, making the sum total of sickness and workdays lost comparable. Why is there no vaccination programme for the common cold?

Alan Bundy
Edinburgh, UK

The lack of a vaccination for the common cold is not for want of trying. Research into both prevention and cure demonstrates how intractable the problem can be, but the matter has become more urgent now that colds have been implicated as a cause of asthma. Smallpox, polio, and influenza, which have been countered by vaccination, are caused by a finite number of viruses. The common cold in contrast is merely a catch-all term for a wide range of opportunist infections of the upper respiratory tract.

Colds are usually ascribed to rhinoviruses which were discovered as causative agents in 1956. However, work at Cardiff University – which has a dedicated common cold research centre – suggests that perhaps only 30 per cent of colds are caused by these viruses. The other culprits can be coronaviruses and adenoviruses, and some colds may in fact be mild attacks of influenza. As for rhinoviruses themselves it is usually stated that there are 99 strains, but the UK National Health Service suggests there are 138 recognised serotypes – and there must be many more waiting to be examined.

Each would require its own vaccine. Furthermore, these small strands of RNA both mutate and exchange genetic information, so there will always be novel strains to deal with. Collectively, cold viruses are among nature's great success stories. A clever pathogen does not destroy its host or immobilise a person to the extent that they cannot walk around and spread their infection. Colds are annoying, and a drain on the economy, but are not seen as enough of a threat to attract the funding given to heart disease and cancer research; nor are they scary enough to make people modify their behaviour to any great extent. Because they are not taken as seriously as, say, Ebola and H5N1, cold viruses are the great survivors.

Christine Warman
Hinderwell, North Yorkshire, UK

Clockwise

Do all climbing plants such as my morning glory grow clockwise and if so why?

Roger Pither
Ottawa, Canada

Your questioner is fortunate in not having bindweed in their garden. A quick look shows that it spirals anticlockwise to the sun (that is when looked at from below), as do all *Convolvulus* species. In contrast, honeysuckle spirals clockwise.

He is less fortunate in not being familiar with the work of the comic musical duo Michael Flanders and Donald Swann, specifically their song 'Misalliance'.

The direction of twist isn't derived from the shoot

following the sun, nor is it different in the southern hemisphere. As the direction of turn is species-specific, it is likely to be genetic. In 2002 botanist Takashi Hashimoto showed that mutations in the genome of *Arabidopsis thaliana* caused clockwise spiral growth in root hairs that are normally straight. Similar mutations could be responsible for opposite helices. The original mutation would have conferred an advantage on climbing plants whichever way they coiled, giving them the ability to climb higher toward the sun on the stems of other plants. If there is a herbicide out there that can prevent this spiralling, I have a garden full of bindweed just waiting for it.

Norman Doidge
Newton Abbot, Devon, UK

Waning woodpile

? Sometimes the wood at the bottom of our woodpile decays to the point where it has only a small fraction of its previous dry mass. What has happened to it? Where did all the carbon go? From the point of view of greenhouse gas emissions, is it better for the wood to be burned in my fire, or for it all to rot back into the earth?

Peter Seligman
Victoria, Australia

Dead wood is first converted into living biomass, mainly through the actions of fungi, other microbes and woodlice. Eventually it ends up as carbon dioxide, methane, water and nitrogen, plus minor quantities of mineral solids, most of which eventually wash away or remain in the soil as fertiliser.

However, in terms of greenhouse gas emissions rotting offers advantages, despite its release of methane. Fire abruptly converts almost all the carbon into CO_2, whereas detritus feeders and agents of decay leave some combustible materials such as lignin and humic acids that act as buffer stores for carbon. Those stores last for years, accumulating in the soil as solids rather than in the air as CO_2.

It would take the CO_2 from burning a long time to settle down innocuously like the trapped residues produced by the wood rotting. There is no simple global limit to how much carbon such storage could accumulate. Its only rivals for sheer magnitude would be the deep-sea accumulation of dissolved CO_2, and carbonate minerals and soils in places such as chalk cliffs and downs.

Jon Richfield
Somerset West, South Africa

Lumber's up

? Can one determine which end of a plank taken from a tree trunk was pointing upwards as it grew?

Ross Kinneir
Bristol, UK

This is often easier than might be expected. The grandstand at Lodge Park on the Sherborne Estate near Cheltenham, UK, was re-floored a few years ago with sweet chestnut (*Castanea sativa*), often referred to as 'poor man's oak'. To minimise waste, the planks were not cut to an even width. Instead, each plank followed the growth of the original tree, which tapered from ground level upwards. The tapering boards were then laid in alternating directions.

In this case, it was easy to tell which end of the plank had been pointing upwards. But even when boards have been squared up, the taper can still be detected in the wood grain, which converges towards what was the top of the tree.

Malcolm Nickolls
Aylesbury, Buckinghamshire, UK

On a macroscopic level the answer is yes, if you have a decent-sized spruce plank. This is because spruce branches always point upwards. Look at the end of the plank first – the curve of the grain will tell you which face was on the outside of the trunk.

Stand your plank up, then find a knot that runs right through it. Knots show where branches emerged from the trunk, so they will run from low down on the inside of the trunk to higher up on the outside.

Adrian Foulds
Glasgow, UK

Above board

? Does wood from the upper part of a horizontal tree branch, which is under tension, have different characteristics to that from the lower part, which is under compression? Is wood ever selected for a purpose on the basis of such differences?

Ross Kinneir
Bristol, UK

In modern times, wood from branches is avoided in most applications because its properties are undesirable. In the past, when there was less demand from builders and steam bending had not been invented, entire branches were commonly used and indeed sought after. Boatbuilders prized curved branches for ribs, and long curving trunks were used for roof framing in buildings such as cathedrals. Branches, often with their trunk attached, were used for wall bracing.

The wood from tree limbs and stems that have not grown purely vertically is called reaction wood. This has different traits in softwoods and hardwoods.

In softwood, there are structural differences on the underside of a bending branch due to compression. The growth rings are wider in this kind of wood, resulting in an eccentric branch with the central pith offset to the upper side. Reaction wood tracheids (elongated cells that make up the core of the stem) are thick-walled and the wood is denser than normal. However, they contain less cellulose, and the cellulose chains are not as parallel to the direction of the cells in the way they usually are. Consequently, the wood is weaker than normal.

The timber also has a greater tendency to split when it is nailed. It accepts stain unevenly and can break without warning. Drying causes the wood to shrink unevenly, resulting in warping. This makes sawing very dangerous because tension can cause the wood to spit back towards the operator, or even explode.

In hardwoods, the abnormal grain forms on the upper side of the stem as it bends and is called tension wood. Because this wood contains a greater amount of cellulose, the timber is often stronger than normal. However, the fibre structure leaves a 'woolliness' (a rough feeling) on the surface when it is machined. Nailing can be problematic because the timber is too hard and nails bend, requiring all nail holes to be drilled out first. Shrinkage and blotchiness when stained are the same as softwood.

The best use for reaction wood is as firewood.

Nina Burdett
Malmsbury, Victoria, Australia

A good example of something that takes advantage of the tension and compression properties of wood is the traditional longbow – although that wood comes from the trunk. The living sapwood stretches, so forms the outside or back of the bow stave, whereas the dead heartwood goes on the inside, or belly, of the stave, and resists compression.

The English longbow was traditionally made from Spanish or Italian yew, which carried less water than English yew. This possibly gave us the posh word for archery – toxophily – which comes from Greek *toxikos* via the Latin word for it, *toxicus.*

Dave Hulme
Stockport, Greater Manchester, UK

Fritillary flitter

? What causes butterflies to flutter and flit up and down in the charming way they do?

Dermot Barrett
Kingston, Ontario, Canada

Apart from flight, butterflies use their wings for temperature control, adjusting themselves to shed or absorb warmth. These actions tend to be conspicuous, which creates selective pressures for species to advertise themselves as vividly as possible to their own species, or warn predators that they are unpleasant to eat.

Such selection favoured flat, exaggeratedly conspicuous wings, which in turn constrains styles of flight, dictating larger, slower flapping movements than the compact, high-frequency wing vibrations of their ancestral moths.

The visual impression that butterflies' style of flight conveys is important; they emphasise it by up-and-down flapping of their billboards, and by individual styles of aerial dancing. This advertises their quality as mates, and provides recognition clues to their kind. For example, long-distance migratory butterflies such as monarchs don't fly like gliders in jungle clearings, or like clouded yellows dancing and dipping over alfalfa fields to attract mates, or leisurely *Acraea* species that only stupid birds taste twice.

Jon Richfield
Somerset West, South Africa

Grow your own

If we did not import any food into the UK, how large a population could we sustainably support by producing our own? Fresh water is obviously an important consideration too.

Jim Watts
Aberdeen, UK

If the UK were unable to import any food, the immediate consequence would be a massive market-driven change to a rural economy. The vast majority of British agriculture is devoted to producing meat, whether directly or through cereal crops grown for animal feed. But meat production is notoriously inefficient, typically requiring ten times as much farmland as vegetables to produce the same amount of food – so meat would become more expensive.

To support our population, we would have to adopt a mainly vegetarian diet, perhaps enlivened by the odd sliver of mass-produced broiler chicken or factory-farmed fish. The countryside would be given over to vegetable production, with enormous acreages of beans, peas and other pulses. The south-east would probably be swathed in polytunnels.

The irony is that if the UK adopted such a diet, people would live longer, healthier lives, leading to short-term population growth. If we also assume further mechanisation, intensive cultivation, cross-breeding, genetic manipulation and whatever extra innovations we come up with, within thirty years we could probably support a population around ten times the current size – say 500 million people.

A lot of water would be required for vegetable production, but the UK is one of the best-watered countries of its size, with frequent rainfall. The problem is ageing infrastructure that fails to collect and transport enough water from the wet north and west, and instead insists on pumping streams dry to support a rapidly increasing population in the south-east. Reorganising the UK water-supply network would be expensive, but if the alternative was people starving, the money would be found.

Nigel Palmer
London, UK

A previous response repeats the old canard about plant-based versus animal-based foods. It is time someone properly examined the relative productivity of different whole-year systems for biomass production and utilisation in the UK.

I have done this for New Zealand (a not dissimilar agro-ecosystem) and have found that while cereals are good for ensuring a high number of people are fed per hectare, they are very inefficient compared with dairy production at meeting protein needs. Legumes are no help. Meat or dairy production are the only way to utilise the biomass we can produce between the grain harvest and sowing the next crop.

When the unsubsidised cost of providing one's daily energy requirements is considered, bread does a reasonable job – about the same as butter. However, cheese is best for meeting protein – or more exactly, amino acid – needs.

Graeme Coles
Coalgate, New Zealand

Pieces of eight

? During a school lesson outdoors, we stumbled upon what we think is an eight-headed daisy. What could have caused this unusual flower to form?

Toby Skinner and Jonathon Chappell
Barton Court Grammar School, Canterbury, Kent, UK

The tip of a shoot, known as the apical meristem, consists of undifferentiated cells with the potential to produce stems, leaves and flowers from a single growing point. When the pattern of growth is disrupted, the growing point can give rise to multiple fused stems. This is called fasciation (a word that comes from the Latin *fasces*: a bundle of sticks), and your eight-headed daisy is a very fine example of the process.

It seems that just the flower stem has become fasciated in this case, but it is possible for entire plants to be affected, or for trees and shrubs to develop flat, paddle-shaped branches with dense clusters of leaves at the edges and tips.

This disrupted growth can be caused by physical damage as a result of insect attack, fungi, bacteria and poor growing conditions, but the main cause seems to be genetic mutation. A lot of media excitement was generated recently by a photograph showing double-headed Shasta daisies growing near the damaged Fukushima nuclear facility. Radiation might conceivably have caused this mutation, but such flowers are by no means uncommon.

The condition can be inherited, and in the UK I have seen numerous double-headed dandelions growing close together, probably the progeny of a single plant. Indeed, the variety of amaranth known as cockscomb (*Celosia*) has been selectively bred for its gaudy ruffled flower heads. Another example of a plant with the mutation is the fantail willow, coveted by Japanese flower arrangers for its twisted, blade-like branches.

The phenomenon occurs in a wide variety of vascular plants including ferns and cacti, and there is nothing sinister about it. You may spot such plants anywhere in the UK and elsewhere. In fact, I cherish an ancient, flattened and twisted wallflower, which survives in the driest spot in my garden.

Chris Warman
Hinderwell, North Yorkshire, UK

Pine fresh

 Just what is that Christmas tree smell?

Thomas Tobin
London, UK

That Christmas tree smell is the scent of coniferous evolution. Over millions of years, these trees have equipped themselves with a cocktail of chemical weaponry, including substances that act as fungicides and bactericides that deter herbivorous pests, large and small.

The chemical combination varies with tree species, but generally consists of a mix of aromatics, including terpenes such as alpha- and beta-pinene, limonene and camphene; and esters such as bornyl acetate. By lucky happenstance, we tend to find these scents appealing – so much so that they are added to perfumes and air fresheners.

If you happen to have an artificial tree, you will smell a different kind of ester, probably a phthalate or suchlike, used to make the plastic fronds softer and more flexible. If so, soak a few tree decorations in pine-scented disinfectant to give a conifer fragrance to your holiday season.

David Muir
Science Department, Portobello High School, Edinburgh, UK

6 The World at Large

Ray bands

[?] I am an Australian merchant naval officer and in the Southern Ocean we see some spectacular sights, but on 22 June 2014 at sunset I saw a phenomenon that I have rarely seen. Rays of light were emanating from the horizon at 180 degrees from the setting sun. Can anyone explain this quirk of nature?

Martin Skipper, Chief Officer, MV Seahorse Standard
New South Wales, Australia

The phenomenon seen by your correspondent is not rays of light emanating from the horizon, but the reverse: rays of shadow apparently converging. They are called anti-crepuscular rays, and are the result of the setting sun – behind the observer – casting shadows from clouds. We are all familiar with the sight of the sun illuminating broken clouds from behind resulting in characteristic sunbeams. These sunbeams are parallel and only appear to spread out because of our perspective from the ground. The process is identical to the way parallel railway tracks appear to converge as they recede into the distance.

Close to sunset, the light rays and cloud shadows will be cast almost parallel to Earth's surface, so can continue for many miles. Dust or mist in the air can reveal the path of light and shadow disappearing into the distance behind us, converging on a point exactly opposite the sun. This is the anti-solar point. At sea level, the phenomenon will only be seen at sunrise or sunset. However, if you are high enough, your anti-solar point will be above the visible horizon, giving you longer to experience this effect. So mountaineers and pilots like me will have far more chance to see them. Mountaineers occasionally have a shadow display of their mountain range cast behind them while the sun is low in the sky.

Martin Powell
Westgate-on-Sea, Kent, UK

Hazy blaze

The worst sunburn I ever received was on a beach in Wales on a dull misty day. I have holidayed many times on Greek islands in the height of summer, but have never experienced sunburn like it. What could be the cause?

Neil Macnaughtan
Edinburgh, UK

I doubt that your questioner used sunscreen on the Welsh beach through being lulled into a false sense of security. After all, the sun wasn't visible and the low temperature would have meant there was less of a sense of burning. On the Greek islands, however, I am guessing that they would have applied sunscreen liberally and subsequently retreated to a suitable area of shade when too hot.

Unfortunately for the skin, the ultraviolet (UV) radiation that led to the sunburn passed through the clouds. And the mist would have acted as a diffuser to ensure an even sunburn. To make matters worse, the sand and sea would have acted as mirrors, reflecting some of this radiation back up – possibly onto skin unused to receiving much sunlight. This was certainly my experience on a visit to Jungfraujoch in the Swiss Alps: after less than an hour outside I had sunburn above my boots where sunlight had reflected off the snow and passed under the hem of my trouser legs.

UV radiation comes in three bands. UVA is the lowest frequency, with UVC being the highest and most dangerous, but fortunately it is blocked by ozone. The one that causes sunburn is UVB light. In 2004, Australian researchers reported an increase of 40 per cent in the intensity of UVB radiation under broken clouds. Although counter-intuitive, this is just one of several studies suggesting that clouds enhance UV. Although the mechanism isn't clear, the effect appears to be maximised when high-altitude cirrus clouds refract UV, and this light is reflected by low-altitude cumulus clouds. Hazy conditions accentuate the outcome. UV-induced damage to DNA triggers the production of melanin, a photoprotective pigment that acts as a natural sunscreen to reduce further damage.

Burning happens through high levels of exposure. Exposure to UVB depends on several factors: the time of year (it is higher in summer); the time of day (with a peak at solar noon when the sun is highest in the sky); altitude (UV levels increase by about 10 per cent for every 1,000-metre rise in altitude); the amount of time spent outside; and latitude. All other things being equal, one would expect greater exposure in Greece, which has an ultraviolet index of three compared with two for Wales. Perhaps surprisingly, sunlight during the summer solstice is only about 20 per cent more intense in Greece than in Wales. But this can be trumped by cloud enhancement of UV or if sunscreen is not applied in chillier climes by the mistaken belief that sunburn is only a hazard on a sunny day.

Mike Follows
Sutton Coldfield, West Midlands, UK

Hue goes there?

? How far beyond the visible spectrum does a rainbow extend?

Greg Parker
Edinburgh, UK

To form, rainbows need light and something transparent to refract it. However, water is only transparent for a narrow sliver of the electromagnetic spectrum centred on visible light. On Earth, only a little infrared and ultraviolet light from the sun sneaks through. But infrared rainbows may form on Saturn's moon Titan, where the atmosphere is transparent to these wavelengths.

Titan is too cold for liquid water to form on its surface. Whereas we have a water cycle on Earth, Titan has a methane cycle, and because liquid methane is also transparent, methane rain might give rise to rainbows. They would be slightly bigger than terrestrial ones with a primary radius of about 49 degrees instead of 42.5 degrees, and would have a hint of orange from Titan's atmosphere. But they would be hard for our eyes to spot, as hazy skies would allow very little visible light to reach the surface. In contrast, the infrared portion of the rainbow would be intense, although you would need special equipment to see it. The bow would have a radius of a little over 50 degrees.

Mike Follows
Sutton Coldfield, West Midlands, UK

The visible spectrum runs from about 400 nanometres (violet) to 700 nanometres (far red). The spectrum produced by a rainbow will depend on two factors: absorption by water, mainly in the form of vapour (because light must pass through water droplets to form a rainbow) and the light source, in this case direct sunlight. Water transmits light best at 400 nanometres. That is why everything looks blue underwater – the longer wavelengths are being absorbed. Water transmits reasonably well in the ultraviolet, up to 200 nanometres, so the strength of the violet end of a rainbow depends on the light source. As it happens, ultraviolet light from the sun is absorbed by the ozone layer and then dispersed in the atmos-phere by what's known as Rayleigh scattering. So direct sunlight reaching Earth is relatively low in ultraviolet, with essentially nothing coming through below 300 nanometres. This then marks the short end of the rainbow's spectrum. The infrared end of the rainbow is more to do with fading out. As light's wavelength increases from 700 to 1,000 nano-metres, the proportion transmitted by water drops by 90 per cent. The available light from the sun is also down to half of its peak at 1,000 nanometres, so we can take that as a prac-tical upper limit. There will be a couple of dips in the spectrum before that, appearing as dim bands in the rainbow. These are the result of absorption by atmospheric oxygen (at 762 nanometres) and water vapour (approximately 900 nano-metres). In practice this gives us a rainbow spectrum ranging from 300 to 1,000 nanometres, although the infrared end will be quite faint.

Guy Cox
St Albans, New South Wales, Australia

Although not directly relevant to your question, it's worth examining a little known detail of the atmosphere's effect on light. We often hear that Rayleigh scattering of light by water vapour makes the sky blue, but the major contributor is actually absorption of light by ozone in the very broad 'Chappuis band' centred in the orange. The colour of the clear sky around twilight is dominated by ozone. Without it, the zenith sky would be a greyish straw-yellow instead of the deep steely blue we see.

Robert Fosbury, Emeritus Astronomer
European Southern Observatory, Garching, Germany

Sunshine down under

I read in *New Scientist* that exposure to the sun in winter at latitudes higher than about 35° results in negligible vitamin D production. For many years here in Tasmania (40° south), I have been exercising naked in the sun just after midday to dose up on vitamin D. But have I been wasting my time in winter?

Guy Burns
Devonport, Tasmania, Australia

I live in Oregon (45° north) and have been measuring ultra-violet radiation for several years. During the winter months, a clear day can have as much UVB radiation as 100 microwatts per square centimetre at sea level.

This is between 25 and 30 per cent of the UVB we receive in midsummer and about 20 per cent of the levels seen in the tropics. For a person with fair skin, adequate vitamin D is produced by fifteen minutes of summer exposure per day, or an hour during midwinter. Those with more melanin in their skin need more sun exposure.

However, the questioner should make sure he continues his exposure at midday, because UVB levels drop off within two hours on either side of the highest sun angle. Of course, cloud cover, rain and snow all severely reduce or eliminate UVB – but then, they also make it hard to exercise naked.

Stephen Johnson
Eugene, Oregon, US

Polar trader

? If I bought 1,000 tonnes of gold bullion in Antarctica and sold it in Mexico at the same price per kilogram, would I make a loss? Surely the consignment would be lighter at the equator than the poles because it is spinning faster?

Richard Byrne
Swanmore, Hampshire, UK

Gravitational acceleration does indeed vary slightly over Earth's surface, partly through a combination of the planet being oblate (not perfectly spherical) and variations in the forces alluded to in the question. There are also more local-ised contrasts caused by differences in rock density and large-scale topographical features – the sea level around Greenland for example is higher than it would otherwise be, because of the large mass of water contained in glaciers. Further minor short-term fluctuations in apparent gravitational acceleration occur because of the relative motion of the sun and moon.

The Earth's spin reduces this acceleration by about 0.3 per cent at the equator compared with the poles. On top of this, the bulge of the planet at the equator means that objects here are further from Earth's centre than they are at the poles, giving another reduction of 0.2 per cent. An object will therefore weigh about 0.5 per cent more at the poles. Mexico City is at an elevation of more than 2,000 metres, which pushes it still further from Earth's centre. However, any additional effect is cancelled out by the fact that Mexico is still some distance from the equator. We can thus estimate that 1,000 tonnes of gold at the poles would reduce to a weight of 995 tonnes at the equator.

So in theory, a sizeable profit could be made if you were to buy at the equator and sell at the poles. However, in practice, the cost of transportation and security would be prohibitive. And the laws of supply and demand suggest that you wouldn't get a good price when penguins or polar bears are the only customers. In addition, gold bullion is usually cast in ingots of a certified mass – so unless you could persuade your customer to buy on the basis of the values on your own scales you would be in trouble.

Simon Iveson
School of Engineering, University of Newcastle, New South Wales, Australia

Getting off the ground

? What is holding back the development of
geothermal energy? There must be an awful lot of
energy down there.

Robert Watts
Bristol, UK

Geothermal energy for heating makes a great deal of sense
in cold climates, and localised 'district heating' systems that
use it have been around for centuries. If you want to
generate electricity, then it's best to find the hottest water
you can. Very hot water is easily accessible in volcanic
regions, but in most areas drilling and fracking are needed
to access the heat deep in Earth's crust. This becomes expensive, especially if you want water hot enough to drive
state-of-the-art power plants that use steam at 600°C.
Typically, water from the ground is much cooler than the
steam we generate from burning coal, oil or gas. Geothermal
power will therefore always be expensive compared with
fossil fuels, but perhaps carbon dioxide emissions will drive
those out of fashion.

Hugh Hunt
Trinity College, Cambridge University, UK

Map lag

? On ancient maps, India is usually portrayed as much
smaller than it is nowadays. Sometimes it's perhaps
only half the size. Surely mariners of the time could judge

distances, especially along a relatively even coastline.
What is the reason for the discrepancy?

T. L. Threlfall
Essex, UK

The root cause of India's varying size is that you can't peel
the surface off a sphere and lay it flat without distorting it.
Imagine trying to do so with orange peel. You have to choose
how to distort it: you can preserve area, distance or direction,
but not all of them. This compromise is inherent in all map
projections of the earth's surface.

There is no perfect depiction, and choosing the appro-
priate one depends on the map's purpose. The sixteenth-
century Mercator projection preserves compass directions:
the north–south and east–west lines are straight (although
all other straight journeys look curved, as you may have
seen on in-flight aircraft route maps). This preservation
of compass direction, as well as the accurate depiction of
coastal features, is what sailors cared about – and they were
the most important customers for maps at the time. But
the Mercator projection has a huge drawback. It makes a
40-kilometre circle around the North Pole as wide as the
40,000-kilometre equator. Africa looks smaller than
Greenland when it is actually fourteen times larger, and
India looks tiny.

So an 1855 alternative called the Gall projection was
revived in 1973 as the Peters projection. This squashes the
vertical distance near the poles to make up for the inherent
horizontal expansion. The result is that northern countries
(and Australia) are unrecognisable, but the relative area of
each country is conserved. If you are prepared to give up on

rectangular maps, a semi-oval is a good compromise, although things at the edges still suffer: the 1805 Mollweide map is a popular equal-area projection, and the National Geographic Society uses the 1921 Winkel tripel projection.

Online maps such as Google Maps still use a simplified Mercator projection because preserving compass directions on a rectangle is their main purpose.

Ron Dippold
San Diego, California, US

Deckchair port

? I notice that Heathrow Airport often tops the list of highest daily temperatures in summer in the UK. Why is this?

Katriona Sewell
London, UK

The south-east of England is an excellent candidate for containing the warmest place in the UK. Not only is it in the south, but it is also on the front line as regards exposure to the warm southerly or south-easterly continental winds that blow during a hot spell.

Heathrow has an additional advantage: the urban heat island effect. Most cities are warmer than the surrounding countryside because their technology-rich environment generates waste heat. Heavily tarmacked surfaces are also better than vegetation at absorbing heat energy from the sun. So it is not surprising that weather stations in London often register the highest temperatures.

However, I don't think Heathrow is necessarily more likely to experience heat extremes than other London airports; it's just that we tend to notice when it does. In fact, the highest UK temperature to date (38.5°C, on 10 August 2003) wasn't recorded in London but at Faversham, Kent.

William Torgerson
Durham, County Durham, UK

Chromatic quandary

> **?** If subatomic particles could be seen with the naked eye, what colours would they be?

Michael Green
Seaview, Isle of Wight, UK

When we look at things, we see light coming from their surfaces. There are two sources for this light. For most everyday objects we see reflected light.

If the light falling on the object is white, then the colour comes from the parts of the visible spectrum not absorbed by the surface. So a 'blue' object is one that has absorbed all the other colours, leaving only the blue component to be reflected. If there were no blue component in the light source, then its surface would appear black instead.

But light can also be emitted by the object itself. All objects at temperatures above absolute zero are constantly emitting electromagnetic radiation.

At room temperatures, this light is at a wavelength below the visible spectrum so we cannot see it with our naked eye (although we can see it using infrared goggles). However, when an object becomes hot enough, its electromagnetic radiation enters the visible spectrum. It starts at the lower red end (for example, a hot glowing coal), then more colours are added as the object gets hotter, until eventually it glows 'white hot'.

If you have an isolated subatomic particle such as a lone electron, proton or neutron, then light waves in the visible spectrum pass right by it without interaction because the particle is so small. So there would be no reflected light to give the particle any colour you could see. It would be invisible, just like the oxygen and nitrogen atoms in the air you are looking through to see this page.

The only way to see such a subatomic particle with the naked eye would be if it were hot enough to emit electromagnetic radiation in the visible spectrum – so its colour would depend on the temperature of the particle.

There is an everyday example of this. Our sun is a ball of plasma made up mostly of free protons and electrons that emit a white light. Other stars have lower surface temperatures, so their protons and electrons look red, while other stars are hotter and have a bluish tinge.

Simon Iveson
Callaghan, New South Wales, Australia

To see atoms as they are, we would need light in the gamma ray range, millions of times shorter than visible light wavelengths; and that implies horrendously high-energy photons. Such a photon striking a particle, or emitted from one, would deflect it like a hammer blow. That is why gamma radiation is so dangerous to living cells; it smashes electrons off atoms. In such light, atoms are gamma ray colour: invisible, because the wavelengths are vastly shorter than any colour we can see.

Jon Richfield
Somerset West, South Africa

I can confidently assert that electrons are green, photons are blue and, if I remember correctly, protons are yellow. I recall discussing this with a colleague in the canteen of a major electronics company when an eminent physicist came over to our table in a fury to tell us that we were talking nonsense. It took some time to explain that we were discussing the colours for the various particles in an animated sequence in a training video.

James Lee
Abergele, Conwy, UK

Under pressure?

? I read in *New Scientist* that the pressure at Earth's core is 3.6 million atmospheres. I would expect the pressure there to be zero, because it arises from gravity, and this must be zero at the core since the pull of the surrounding matter is equal in all directions. Can someone get to the bottom of this?

Shawn Charland
Ottawa, Ontario, Canada

Don't confuse the pressure at any point with the gravity at that point. Certainly the gravitational forces at the centre of mass cancel out, reducing weight to zero. But the gravity at every other point inside Earth is directed towards the centre of mass, increasing the pressure at all deeper points.

Imagine a tightly inflated elastic balloon: the gas at the centre is under high pressure, even though that pressure is applied only around the outside.

Jon Richfield
Somerset West, South Africa

The pressure at the bottom of the Mariana Trench in the Pacific Ocean is more than a thousand atmospheres. If we were able to find a trench elsewhere that was twice as deep, the pressure there would be more than double that.

Keep going down further and the pressure keeps going up. At the very centre of the earth, the pressure is caused by the weight of about 6,400 kilometres of liquid iron and rock lying above.

Just because gravity is zero at Earth's centre doesn't mean that the stuff overhead has suddenly become weightless. Imagine the heavy column of molten rock below Singapore that bears down on the core. On the other side of the globe, the weight of the molten column under Ecuador pushes down in the opposite direction.

An object at the centre of Earth's core may be weightless, but it is like an orange pip being squeezed between your fingers.

Hugh Hunt
Cambridge, UK

Revolving core

? I keep reading about how every so often Earth's north and south magnetic poles 'flip'. How long does it take for this to happen? Moments or years? And does the magnetosphere 'turn off' when this change is taking place? Or is it more like turning a bar magnet through 180 degrees, with the poles moving across the planet's surface and passing through the equator?

Geoffrey Clark
Douglas, Isle of Man

Reversals occur once every 300,000 years on average and it takes a few thousand years for the poles to switch places each time. Although the field diminishes, it never falls to zero. The last reversal took place 780,000 years ago, so perhaps the next one is overdue.

Earth's magnetic field is generated because the planet behaves like a self-exciting dynamo. As the molten outer core freezes to become part of the solid inner core, latent heat is released, which drives convection currents in the outer core. Because the core is metallic, it contains de-localised electrons. The convection currents and Earth's spin cause these electrons to move, creating electrical currents and generating a magnetic field.

The magnetometer – a device that measures the strength and direction of magnetic fields – was only invented in 1832 by Carl Friedrich Gauss. So, in order to gauge Earth's magnetic field longer ago than this, David Gubbins of Leeds University in the UK looked at ships' logs kept by early ocean-going explorers. Navigating using a compass and sextant, these seafarers measured and recorded magnetic declination – the angle between true north and magnetic north. This allowed the strength of Earth's field to be reconstructed and shows that it is about 10 per cent weaker than it was in 1860, suggesting that we might be heading for a reversal. However, this drop in magnetic field may also be due to another unconnected effect.

In a further bid to understand what is going on, Daniel Lathrop and his team at the University of Maryland have constructed a spinning sphere of molten sodium called the three-metre dynamo experiment. They are hoping to show that the spinning sphere acts as a dynamo, generating its own magnetic field. This continues a tradition begun by William Gilbert, who built a 'terrella', a sphere made from a magnetic lodestone, to model Earth's magnetic field. His work, published in *De Magnete* in 1600, was not surpassed for two centuries. Largely forgotten, he pioneered experimental science and greatly influenced Galileo and those who followed.

Mike Follows
Sutton Coldfield, West Midlands, UK

We know of an underground volcanic intrusion that cooled over hundreds or thousands of years, where different parts of the structure have preserved distinctly different magnetic orientations. This has been interpreted as meaning that the duration of a reversal event is, indeed, on the timescale of millennia.

This is in general agreement with our (admittedly inexact) theoretical models of these events. This is why geologists do not give much credence to 'polar-reversal-doom' scenarios; over an individual animal's lifetime, the changing of fields is unlikely to override other cues such as prior experience.

From a different point of view, organisms that depended solely on magnetic fields for migration would have died out at every reversal. This is unlikely – the use of multiple cues is far more likely than reliance on one, significantly variable prompt. If, for example, your magnetic sense disagrees with your 'sunrise-in-the-right-eye' rule for a to-northern-summer migration, then the organisms which gave more weight to one not the other would have had worse migration effectiveness and offspring survival. The ones which blended the two would have survived and bred better.

Aidan Karley
Aberdeen, Scotland, UK

Surf's up

? On a day of meagre waves, surfers often say that one should wait for the tide to turn because the waves will get bigger when the tide is 'on the push'. Is there any reason to believe that waves will increase in height with an incoming tide?

Colin Risner
St Austell, Cornwall, UK

In brief, yes. Surf arises from swells that may have travelled thousands of kilometres. If swells travel over deeper areas, they gain speed, and thereby stretch out and lose height. Conversely, as depth decreases towards the shore, the shoaling water slows the leading zone of each swell. The water behind piles up until it eventually curls into a breaking wave. The faster the wave, the more energy it has, and the more generously it breaks.

Ebbing water moves away from the beach. This reduces the forward velocity of incoming swells relative to the beach, thereby sapping the surf's energy. Incoming tides increase the waves' velocity towards the beach, accordingly increasing the surf's energy. You might reasonably object that the ebbing and flooding of tides is too slow to be noticeable, and for swift waves that is true. But for say, North Sea waves, which are smaller and therefore slower, the relative difference is greater. Kinetic energy increases with the square of the velocity, so in surfing a difference of just 2 kilometres an hour in wave speed can turn stingy ankle-busters into generous nugs.

Jon Richfield
Somerset West, South Africa

7 Physics

Atomic bonds

 On an atomic level, how do Post-it notes stick to things?

Felix Barbour
London, UK

Post-it notes are a classic application of polymer chemistry. Funnily enough, the stickiness in question was discovered when Spencer Silver was trying to produce an incredibly strong adhesive in 1968. Instead, he produced an incredibly weak one, but most discoveries are accidents, so who is going to complain? Perhaps just students like me as we stick up reminders of our exam dates and assignments.

The glue used in Post-it notes is a pressure-sensitive adhesive. This means you only need to apply light pressure to stick the note to something. The bond between the note and the surface is formed through a fine balance between flow and resistance to it. The adhesive can flow just enough to fill tiny crevices on the surface, but will resist flow enough to remain there. This produces the bond between the note and the surface.

If we zoom in further, we can see that on a molecular level, the biggest contributors to the bond strength are van der Waals' forces. These are created when a molecule has more of its electrons on one side than the other, producing a dipole (like a tiny magnet). This induces the opposite dipole in another molecule nearby and the two stick to each other. Van der Waals' forces are normally weak, but they increase in strength as the size of the molecules grows.

Bradley Clarke
Brighton, East Sussex, UK

Post-it adhesive sticks in much the same way as any other pressure-sensitive contact adhesive: through a combination of van der Waals' forces, a decent balance of cohesion (the glue molecules sticking to each other) and adhesion (sticking to everything else), and good contact with the microscopic topography of the target surface.

The more interesting question is how the notes peel off harmlessly. In practice, peeling leaves microscopic traces of Post-it glue on the target surface, but those are generally too small to cause visible damage. The secret is the molecular structure of the polymer, which forms a delicate network that links relatively large lumps together. Imagine it as being like a tennis net with sticky balls attached at intervals over most of its surface. The Post-it paper soaks up and holds powerfully onto one full surface of the adhesive mass (the tennis ball), but only one cheek of each sphere (or ball) sticks out, just touching the other surface.

This has two important effects: ensuring that cohesive forces greatly exceed adhesive ones, and that peeling the paper off the target pulls on only the next ball in line – each requiring only a small force at a concentrated point, almost like undoing a zip tooth by tooth. Strong contact adhesives are different in that they spread the peeling forces over as wide an area as possible. They make it hard to pull any point free without fighting all the neighbouring adhesive material at the same time.

Jon Richfield
Somerset West, South Africa

Coloured dots

> ? I've noticed that when an object is seen from a distance in daylight, the colours appear changed. For example, when I observed my wife walk round a lake in Switzerland, her pink top looked white and her blue trousers looked pink from 600 metres or more away. Do surfaces need to be sufficiently large in our field of view for us to perceive their colours correctly?

Neil Croll
Derby, UK

In the centre of your visual field there is a small area that is effectively blue-blind. This 'foveal tritanopia' means that objects with colours that differ only in how much blue they contain become indistinguishable when they are very small – about the size of a tennis ball viewed from the other end of the court. So white and yellow will look identical, as will red and magenta, or blue and black.

This phenomenon has been known empirically for a long time. Naval signalling flags are designed so they cannot be confused even when viewed at a distance where this effect could manifest. Similarly, the rule of tincture in heraldry forbids a yellow emblem on a white background or vice versa, or blue on black, and so forth.

Roger Carpenter, Professor of Oculomotor Physiology
University of Cambridge, UK

This effect is called aerial perspective and it is an important technique in a landscape painter's toolbox. Diluting colour intensity by blending it with white mimics the effect of the atmosphere on distant objects. It is the reason far-away hills have a bluish or purple tinge, and why the colours on your wife's clothes appeared pale. In a landscape painting, bright colours such as reds and yellows are best used in the foreground, whereas pale blues and other diluted colours will give the illusion of depth. Of course, the Fauvists broke all these rules, favouring strong colours over realism, with wonderful results.

Ingrid Banwell
Sydney, New South Wales, Australia

Thin iron line

? When you demonstrate the magnetic field of a bar magnet using iron filings the filings form lines. But isn't the field a continuous plane? What makes the lines form and why do they spread apart at the sides of the magnet and converge at the poles?

R. Milton
Hoo, Kent, UK

A magnetic field permeates the materials and space around it. The field is in 3D and it is continuous, but we often visualise it using field lines in 2D. Field lines are not a physical reality, but serve as a handy way to indicate field direction and strength. When iron filings are scattered on paper over a bar magnet, each filing becomes a temporary mini-magnet and aligns with the field. All the lines of filings have an induced field opposite to that of the magnet, so the lines mutually repel, causing them to space themselves out.

The magnet's field is weaker at the sides, so the mutual repulsion is more apparent there than at the poles where the magnet overpowers the filings' mutual repulsion, drawing the lines closer together. It's easy to demonstrate this mutual repulsion by attaching two 5-centimetre nails pointy ends up to the bottom of a bar magnet held vertically. Gently move the magnet to try to get the nails to touch. The motion can be mesmerising.

David Muir
Science Department, Portobello High School, Edinburgh, UK

The field is indeed smooth – until you put iron filings into it. These become tiny bar magnets, their poles opposite to the field. This twists them and aligns them with the field, then makes them join nose-to-tail in the way bar magnets like to join. But bar magnets don't like to be side by side, so they repel their neighbours and the strings of filings push each other apart.

Another way to look at it is that by opposing the field that magnetises them, the iron filings reduce the local magnetic field and end up being pushed to areas where the residual field is strongest. The weakening in the lines seen at the sides is the natural shape of a dipole field. Along the magnet's axis, the fields of all the domains – the little magnets that we can imagine the bar magnet to be made of – sum together making the field strong. As you go off axis, away from one pole, it weakens, dipping to a minimum alongside the magnet. It builds up again as you approach the opposite pole.

Alec Cawley
Newbury, Berkshire, UK

Rules of attraction

? In my business, we use rare-earth magnets that are about 32 millimetres across and 8 millimetres thick. Four of them will securely hold a 1-kilogram device on a bulldozer while it is operating. But when a delivery of 1,000 magnets arrives, the package has nearly no magnetic field around it. Why is that?

Chris Seymour
Lota, Queensland, Australia

When a package of 1,000 magnets gets shipped, the magnets are arranged into cylinders. One cylinder will consist of these 32-millimetre by 8-millimetre magnets stacked with the north pole of one facing the south pole of the next, as it is simply impossible to stack them any other way. A stack of magnets is no stronger than a single magnet. Next, the cylindrical stacks will be packed in a box with the polarities of the stacks in alternating directions. Thus, the end of the package will have north and south poles alternating like the squares on a chessboard.

The reason for doing this is to have the magnetic effect of the alternating stacks largely cancel each other out. It would be hazardous to ship the magnets without doing this. The ends of the package will still exhibit some magnetism, but once you add in a thickness of packing material, there really is only a very weak field on the outside of the package.

David Emanuel
Tulsa, Oklahoma, US

Bubble trouble

[?] Why is it that bubbles released at some depth in water appear to take on the shape of mushrooms as they rise?

Patrick Casement
London UK

Bubbles can take on a number of different shapes depending on their size, velocity and the purity of the water. The boundary between air and water has surface tension, so a static bubble will take on a spherical shape to minimise its surface area. But as it moves through water, the influence of drag forces comes into play.

You might expect bubbles to adopt a streamlined aerofoil shape. However, the opposite is true. According to Bernoulli's principle, the surrounding fluid must accelerate as it passes the bubble, which results in a pressure drop around its equator. The bubble then expands so those larger than about 1 millimetre in size become noticeably disc-shaped as they rise.

As bubbles increase in size, turbulence causes them to oscillate, creating extra drag and slowing them down. Above about 20 millimetres diameter the increased turbulence causes bubbles to enter what is officially called the spherical cap regime – the mushroom shape referred to in the question. The upper half of the bubble is a hemisphere, but the turbulent vortices at the rear cause the underside to become a chaotic mass of fine bubbles breaking and coalescing, and these get caught in the wake behind the rising spherical cap.

The pressure acting on a bubble decreases nearer the surface, hence bubbles expand as they rise, making them more likely to enter the spherical cap regime. Greater depth also provides a greater distance over which rising bubbles can collide and coalesce, as large faster-rising bubbles tend to catch up and engulf smaller ones above them. This process can continue until they reach the critical size for spherical cap behaviour.

Simon Iveson
Callaghan, New South Wales, Australia

Light fantastic

? I have a luminous watch that I wear at night. Even though I put it on a south-facing window ledge to refresh it during the day, the dial has always been very faint. One night, finding it impossible to read the time, I flashed an LED bicycle light at it for a second. The watch immediately became very bright. How can one second of torchlight have more effect than a whole day of sunlight?

P. J. Stewart
Oxford, UK

Most glowing watch dials are coated with a paint containing a phosphorescent substrate such as strontium aluminate. When electromagnetic radiation strikes this, some incoming photons will have just the right amount of oomph to kick electrons in the atoms of the substrate up to higher energy levels. These excited electrons then fall back to a calmer state, emitting a new photon in the process.

Whereas the energy exchange in fluorescent materials is almost instant, phosphorescent materials can tie up the photon's energy in a 'triplet state' that can take minutes or even hours to decay. The result is that phosphorescent materials continue to emit low-intensity light for a considerable time after the initial radiation charge.

The average intensity of solar radiation on Earth is about 1.37 kilowatts per square metre: the solar constant. So every second, about 1 joule of energy would hit a luminous watch face of 7 square centimetres kept in constant sunlight. However, if it remains facing south throughout the day in the northern hemisphere, it will not be in constant direct sunlight and will receive much less energy.

Assuming the writer observes the dial soon after sundown, it is faint because the westerly setting sun has only indirectly, and weakly, illuminated the south-facing watch. In any case, the dial's energy is continuously leaking away as electrons settle into lower energy levels.

By contrast, an LED bicycle light has a typical power rating of 3 watts. If flashed on a watch face at close range for a second, practically all of the 3 joules of energy emitted will be directed at the face and the dial will glow brightly.

White LED light not only has more energy than direct sunlight, it also tends to be more blue-shifted, so it gives off higher-energy photons that are more likely to cause electron excitation in the dial. By contrast, sunlight consists principally of lower-energy yellow and infrared light.

However, the dial's luminosity would almost certainly persist for longer after three hours of afternoon sunlight, say, than after a blast of LED light. This is because increasing the duration of irradiation increases the probability that each electron in the phosphor will be hit by a photon, so more triplet states can potentially be set up.

Sam Buckton
Chipperfield, Hertfordshire, UK

In the past, the paint on luminous watches contained a radioactive material and a phosphor. The radiation continuously excited the phosphor, so it would glow with the same intensity all night. As a schoolboy, I put a Geiger counter to my watch and was surprised at how strongly it reacted.

Now, luminous watches contain a long-persistence phosphor that is excited when exposed to light. This lasts a while but decays exponentially, and is only really bright soon after excitation. So a whole day's exposure to daylight counts for little – it's only the last few minutes that matter.

I was outdoors most of the afternoon, and half an hour after sunset (in the Australian winter) the luminescence was invisible. A quick blast with an LED torch made it bright again. High-power white LEDs emit a lot of ultraviolet light, and this is efficient at exciting the phosphor.

Incandescent lamps work too, but are less effective.

Guy Cox
Australian Centre for Microscopy and Microanalysis, University of Sydney, Australia

Watch manufacturers use different phosphorescent materials on their luminous dials. Some glow brightly after exposure to light but fade to invisibility in a matter of hours or less. Other materials are not so bright but last longer. I have a watch with a dial that is still discernibly luminous after fourteen days.

Exposing the watch on a south-facing window ledge all day could be counterproductive: after the phosphorescent material has become saturated, further exposure adds nothing to its performance, but sustained exposure to solar rays will age the material.

Rather than leaving the watch to age in the sun, giving it

a blast from an LED torch at bedtime might produce enough phosphorescence to last through the night. Looking slightly to one side of the watch might also help. There is a higher concentration of rod cells away from the eye's centre of vision, and these function at low light levels.

Terence Collins
Harrogate, North Yorkshire, UK

Time for tee

? When a golfer hits the ball off a tee, the tee normally remains stuck in the ground or is projected a few feet forward. But, occasionally, the tee is found several metres directly behind the player. What is the mechanical explanation for this strange behaviour?

Peter Bower
Moffat, Dumfriesshire, UK

This occasionally happens when using a driver. The tee itself is up to about 9 centimetres long, usually made of wood or plastic, with a hollow conical top which tapers down to a point. The ball sits on the top with the point inserted vertically 2 or 3 centimetres into the turf.

When played correctly, the ball will be swept off the top of the tee, with the bottom of the club also striking the top few millimetres of the tee. At impact, the direction of the club head will be approaching the bottom of its arc but critically – if the tee is to be propelled backwards – still in a shallow downward direction.

The impact briefly traps the ball in the conical top of the tee, producing both a forward and downward force on the tee. Golfers often refer positively to turf being 'springy'. With the tee set in springy turf with a fibrous damp base, the turf will push back on the tee with an equal and opposite force that pushes upward and backward, courtesy of Newton's third law of motion.

Once the ball and club have gone, this reactive force holds sway. So, if the turf conditions, insertion of the tee peg and angle and direction of impact are just right – known to golfers as 'the Goldilocks zone' – hey presto, the tee goes backwards. More often, the tee stays in the ground, leaning forward and unable to escape the turf, or it spins up out of the ground, falling straight down or slightly forward.

It is often said the backward-flying tee only accompanies good shots, which seems to be backed up anecdotally, but maybe we should have known this because Jack Nicklaus, the greatest-ever golfer, was known as 'The Golden Bear'.

Terry Healy, PGA Professional
Portsmouth Golf Centre, Hampshire, UK

This question occurred to me years ago when I was a keen golfer, so I did some experimentation.

I concluded the primary cause of this variation was the height of the exposed tee, or rather how much of it was in the ground. Most golfers will hit the top of the tee as well as the ball. If the ball is teed high the tee goes forward, if it leaves the ground at all. It is quite loosely fixed in the ground and so takes up the motion of the club.

However, if the ball is teed low it can travel back. Most of the tee is now in the ground so the momentum acquired by the tee when struck is insufficient to pull it out, but it will still be pushed forward. This enlarges its hole, loosening its grip on the soil. Tees are often made of bendy plastic and will rebound backwards and, being now loosely held, will flip out of the ground in this direction. This might not work with old-style wooden tees, which can break.

As for the ball, there are no explanations for its behaviour. I once mishit the ball off the first tee. It struck a concrete marker in front of me, sailed over my head, bounced on the clubhouse roof and landed on the 18th green. I had two putts for a round of 3 . . .

Dave Bickenson
Saint-Mesmin, France

Filtered water

As a photographer I have always wondered why the light reflecting off water foliage and glass is affected by polarising filters, but the light reflecting off metal surfaces is not.

Joe Martinez, via email

The fundamental reason for the difference is the way the free electrons in the metal interact with light waves. Key to understanding this are Fresnel's equations, which describe how much light is reflected by a material based on its refractive index. These calculations depend on the polarisation of light in two planes: perpendicular to the plane of incidence ('s') and parallel to it ('p').

The equations predict that at a particular angle known as Brewster's angle, the amount of 'p' polarised light will fall to a minimum. For glass, transparent plastics, and water the refractive index is between 1.33 and 1.6. This leads to a Brewster's angle of between 50 and 60 degrees, causing the 'p' reflectivity to drop to zero.

So if your filter blocks the 's' wave, you will see no reflection at all. For metals, it's more complicated. The refractive index becomes a 'complex' number with real and imaginary parts. The imaginary part describes how strongly the metal absorbs light and is a very large number. When plugged in to Fresnel's equations, this also makes the Brewster's angle large, up to around 85 degrees. The reflection of the 'p' polarisation only dips a little, maybe to 74 per cent, so the filter wouldn't produce a strong contrast.

Foliage of course is a complex mix of materials. It contains a lot of water, doesn't have many free electrons, but it does absorb light. Its refractive index isn't easy to calculate, but will be between that of water and a metal. It would have a significant imaginary part to the index, but much smaller than for a metal. So one would expect the effect of the filter to be weaker on leaves.

Harvey Rutt
Southampton, UK

Furless midriff

My Year 7 class left brine in evaporating dishes next to a sunny window. Several pupils ended up with dishes where there were salt crystals at the bottom, a rim thickly furred with salt, but nothing in between. Why is this?

Mr Holden and 7Y2
St Thomas More School, Crewe, Cheshire, UK

There is actually a thin layer of salt deposited in the mid-band of the dish. The real question is why a much thicker layer of crystals formed further away from the liquid. This is because of the combined effects of capillary suction and evaporation. Around the edge of the liquid, surface tension causes the solution to rise up the wall slightly. This layer is thin and has a high surface area to volume ratio, so experiences a high rate of evaporation that causes salt crystals to precipitate around the edge.

These make the surface rougher, enabling capillary forces to suck solution further up the dish's walls. The further it moves from the main reservoir, the more exposed it is to evaporation and the less opportunity there is for the high salt concentration to diffuse back into the bulk. The solution thus becomes supersaturated more rapidly at the rim, and the process of precipitation accelerates.

Eventually the liquid is sucked so far from the bulk that the rate of evaporation equals the maximum capillary flow rate, at which point most of the salt crystallises out and forms the large deposits you can see. In some situations with very low evaporation rates, I have seen salt deposits continue to grow over the rim down the outside of the dish and onto the bench.

Simon Iveson
Callaghan, New South Wales, Australia

If some of the evaporation dishes are slightly porous, the dissolved salt could be migrating through the clay. The water evaporates preferentially at the highest point, leaving behind the precipitated salt on the rim. The middle is too dilute and has too low a concentration of solution to form crystals.

As a potter I see this when I reclaim clay slop in porous bowls in which dissolved salts in the clay precipitate at the top lip only. Egyptians used salt migration to make self-glazing pots. Salts would migrate to the outside of the clay and would then combine with the silica in the clay to form a glaze when firing.

Today, potters add soluble forms of sodium to trap carbon in Shino glazes to achieve a charcoal-grey spotting pattern. The sodium migrates with the water from the glaze to the outside and top of the pot. The sodium compounds melt early in firing, and if there is a lot of soot in the kiln chamber at that temperature the soot will get trapped in the glaze appearing as black areas on the finished pot.

Jennifer Assinck
Newmarket, Ontario, Canada

Toil and bubble

? My daughter was blowing bubbles from a pot she bought from a toyshop. We were surprised to see that the bubbles varied in colour. Some were blue, some green, some red or pink or orange. And sometimes the outer skin of the bubble appeared to be one colour and the inside another. At first we thought it must be caused by their size or their angle to the sun. But it wasn't. Small or large bubbles could be the same colour and the colour remained the same wherever the bubbles drifted. Different colours would come from a single blow of the soap substance used to create them. So we decided it must be down to the chemistry of the soap. Can anybody explain further?

Keith and Joanna Hamley
Hastings, Sussex, UK

The physics of soap bubbles is a fascinating subject. The colours are most commonly caused by thin-film interference between light rays reflected from the outer and inner surfaces of the bubble. Depending on the thickness of the bubble wall, certain wavelengths of light will interfere constructively, giving rise to strong colours. As the bubble evaporates, the wall thickness will change and hence so will the colours.

Because the soapy water tends to flow downwards under gravity, the thickness of the bubble wall may also vary from top to bottom, giving rise to horizontal bands of colour. This can be reduced by thickening the soap solution, for example, by adding glycerol. When the bubble wall becomes very thin – thinner than the wavelength of visible light – the colour will disappear. These regions are transparent, but may appear dark because they are usually surrounded by coloured ones. The bubble will burst a few seconds after these patches appear.

Colours can also be produced by adding dyes to the soap solution. Most water-soluble dyes will not work, however, because of the thinness of the bubble walls, and because the water collects at the bottom of the bubble. The dyes need to bind to the soap used to make the solution. A local ice rink has regular 'kiddy club' sessions for toddlers, and they very often make soap bubbles for the children. The cold air and high humidity allow the bubbles to persist for up to a minute and they show quite spectacular colours.

Bill Tango
Manly, New South Wales, Australia

Wibbly wobbly

? Lots of people have seen a video of wobbly lamp posts on the M62 motorway in northern England (*bit.ly/M62LampPosts*). The posts are swaying erratically in the windy weather and news articles cited vortex shedding as the reason. What does that mean?

Joyce O'Hare
Cleckheaton, West Yorkshire, UK

When a flow of air or gas passes around a vertical obstruction such as a lamp post or a chimney, a vortex will form on one side downstream of the obstruction. The vortex will then break away, followed by another vortex forming with an opposite rotation on the other side. The process is repeated at regular intervals.

This shedding of eddies or vortices creates a regular change in pressure on each side of the obstruction, and if this coincides with the natural frequency of oscillation of the obstruction, it will cause the object to vibrate to and fro as in the case of the lamp posts in the video. If the swaying persists and causes the vibration to increase, the structure can fail. In the early 1960s there was a 30-metre steel incinerator chimney at the Distillers Company chemical works at Saltend near Hull in the UK. Despite being stabilised by guy ropes, this swayed alarmingly in fairly gentle 35-kilometre-per-hour winds.

It was so bad that the company took the problem to the National Physical Laboratory. Their solution was to fit helical fins called strakes to the chimney, which caused the eddies to shed at different times at different heights along the length of the chimney, thus preventing any natural vibration from being set up. This spiral design can now be seen on metal chimneys all around the world, and every time I see such a chimney my mind goes back to its progenitor at the Distillers factory where I once worked.

Vortex shedding also causes problems with the vertical

cables of suspension bridges. Because spiral fins are not a practical solution here, the problem is solved by fitting dumb-bell-like weights to calculated positions on the cable, which dampen its natural vibration so that there is no risk of fatigue failure.

Tony Finn
Hedon, East Yorkshire, UK

We are immersed in vortices, but very rarely see them. They are the reason why we can hear the wind sighing through pine trees, why leaves flutter, and why flags flap. Yacht crews can see them in the flapping of a loose sail, hear them in the hum of the rigging, and are annoyed by them when the ropes bang against the mast at night. Overhead power or telephone wires and even wires in fences can sometimes be heard to 'sing' when the wind is blowing across them. This is obviously difficult to hear in a noisy city but can be heard in a quiet country lane.

Vortices can sometimes be seen in a smoothly flowing stream as water passes around a rock or in the air if smoke or dust makes the flow visible. Vortices can form almost anywhere that fluids move, especially in low-viscosity fluids like air and water. Any object that is not streamlined will create or 'shed' a trail of vortices or swirls of fluid on its downstream or downwind side.

Have you ever wondered why a flag being towed by a helicopter does not flap as furiously as one on a pole in a stiff breeze? A flag flaps in the vortices shed by the pole or wire at its leading edge. A large-diameter pole creates large vortices that can ultimately shred a flag, as anybody who has seen an old straggly flag flapping in a strong wind can testify. However, the thin wire that holds a flag beneath a helicopter creates much smaller and faster vortices, causing only ripples in its flag.

Usually vortices form alternately to the left and right of the leading object and each swirl of fluid applies a small force to the source of the vortices. How rapidly they are repeated depends on the size of the object and the speed of the flow. When vortices are synchronised or 'in phase', the forces start to add up and can become destructive – uncontrolled vortices can cause serious damage.

Vortices can also be very useful and are used industrially in a robust method of fluid flow measurement. A body is placed inside a pipe to create vortices. Pressure sensors detect the vortices as alternating pulses and the rate of the pulses is proportional to the flow rate.

Martin Greenwood
Stirling, Western Australia

Caffeine shakes

I am a regular traveller between the Scottish island where I live and the mainland. The vibration on the ferry sets up an interesting interference pattern on the surface of a cup of coffee. It forms quickly after the cup is put down then remains fixed. Presumably waves are reflected from the sides of the cup and subsequent waves reinforce the earlier ones. The pattern remains as the coffee level gets lower. In the interests of research perhaps next time I should investigate whether the effect occurs in a glass of beer! An explanation would be appreciated.

Jack Harrison
Tobermory, Isle of Mull, UK

A standing wave pattern will form in a cup when reflected waves arrive back at their starting point with the same phase as the outgoing waves. These reflected waves then add to the next outgoing wave. On the ferry, if you pick up the cup, the ripples will disappear. This makes it seem as if you absorb the ferry's vibrations, but because you contain a lot of water your body only partly absorbs the vibrations. In reality, they are reflected back into your hand by the rigid cup wall before they make it into the coffee.

Here is an experiment: on your next trip put your forearm on the vibrating table and pick up the coffee. The vibration pattern will disappear. Now with the same hand put one finger in the coffee and you may see the pattern appear again (it is best to wait until the coffee has cooled). The vibrations pass from the table through your finger and into the coffee. For anyone reading this who is not going on a ferry soon I suggest you look at my video on the acoustic properties of cooked spaghetti (*bit.ly/NSshakes*).

Jeremy Hawkes, via email

You don't have to catch the ferry to Tobermory to see circular wave patterns in a cup of coffee. The effect makes for a great little experiment. Find a variable DC power supply, a toy motor and a wheel that fits the motor's axle. Glue a weight such as a small nut to the wheel so that it spins off-balance.

With the motor taped to the side of the cup you should be able to produce the standing wave patterns described. The seemingly stationary ripples are called capillary waves, and are dominated by surface tension effects. They are also known as Faraday waves after Michael Faraday, who first described them in 1831. The pattern of the waves is determined by the frequency of the vibration, so the ripples in the coffee on the ferry can be used to estimate the revolutions of the ship's engine. At higher frequencies, capillary waves form more complex patterns and intricate examples can emerge.

Neil A. Downie, via email

Bathed in heat

? I had a hot bath one evening and decided not to let all the heat go to waste. So I left the plug in until the water had given up all of its heat to the house. But when is the best time to pull the plug out? Should I wait until the water is the same temperature as the ambient indoors air, or should I keep the extra thermal mass in the bathtub, and then pull the plug when the temperature outside the house is at its lowest?

Matt Rutter
Reading, Berkshire, UK

The water serves two functions. It provides a source of heat until it cools to the ambient temperature, and it is also a thermal mass. Assuming the water equilibrates with the surroundings of the dwelling and the heat only leaves the house at a set rate, the additional mass of the full tub will cause the house to cool at a slower rate. But the interior of the house will also take longer to heat up.

The most efficient thing to do is to drain all the water out of the bath at the point when the lowest interior house temperature has been reached – you may need to have a thermometer to hand. Draining the bath at this point will lower the thermal mass and will then allow the interior of the house to heat up faster when the outside temperature increases once more, because heat will not be wasted in increasing the water temperature.

Neil Ayre
Kalgoorlie, Western Australia

Inside out

? What effect would the accelerated particles in the Large Hadron Collider have if they hit a human body? And what would happen if particles aimed at each other collided inside you or collided with one of your own particles?

Mick Johnson
Leeds UK

At least one unfortunate person – Anatoli Bugorski – has already been struck by a high energy particle beam. In 1978 his head was briefly caught in the proton beam of a particle accelerator, the U-70 synchrotron in Russia. He reported seeing a flash of light 'brighter than a 1000 suns' but experienced no immediate pain. The full extent of his injuries only became apparent over the next few days – he lost half of the skin on his face where the beam had burned a path through it, and experienced various other complications.

To the surprise of doctors he survived, albeit with lifelong symptoms. It isn't clear whether the beam was operating at its full capacity of 70 gigaelectronvolts at the time of the accident. Presumably if one did the same thing at the Large Hadron Collider, which has a beam energy of 6.5 teraelectronvolts –

almost 1,000 times greater and double again at the point where the beams collide – the aftermath would not be pleasant.

There is also the experience of Ray Cox, one of the North American victims of the infamous Therac-25 radiation therapy machine in the mid-1980s. As a result of poor hardware and software design, he was targeted with an electron beam that was more than a hundred times the intended dose. In his own words this felt like 'an intense electric shock' and he fled the treatment room screaming in pain. Some victims of the defective equipment died of radiation poisoning. The beam energy of this machine was a mere 25 megaelectronvolts.

Richard Miller
London, UK

Wave goodbye

? Sound waves fade over distance. However, we can see light waves arriving from billions of light years away. Why don't incoming light waves dissipate before they reach us?

Anthony Turner
Auckland, New Zealand

The answer stems from the fact that light energy is not dissipated as it travels through the vacuum of space, whereas sound cannot travel through a vacuum and needs a material in order to propagate.

When sound or light waves pass through material, the matter takes a toll. Even in transparent material, light is absorbed and the amount lost often depends on the light's wavelength. In the ocean, for example, a large fraction of the red end of the visible spectrum dissipates within about 10 metres of the surface – and not much light of any wavelength reaches depths greater than 50 metres even in very clear water. By contrast, sound usually suffers less attenuation through water than light does. Certainly a submerged scuba diver will hear the sound of an engine and rotating propeller even if the boat responsible cannot be seen.

Unless it is absorbed by an object blocking the way, sound or light from a point source spreads out in all directions and its energy expands over the surface of an imaginary sphere which gets bigger with time. So part of the reason why sound (or light) seems to fade with distance from the source is because its energy is being spread over a greater surface area. Of course, if stars and galaxies did not radiate such prodigious amounts of light energy they would not be visible from billions of light years away.

Mike Follows
Sutton Coldfield, West Midlands, UK

A chill wind

? Objects that enter Earth's atmosphere are subject to extreme heat through air friction – so much so that returning spacecraft have to be very well insulated, and rocks and small asteroids burn up. How fast would I have to go to feel heat from air friction? When I'm on my bike going downhill quickly, I can only feel it getting colder and colder.

Ben Cordle
London, UK

I suspect that when cycling downhill, what you feel is the cold from chilly air rushing past you, or cold air pooled in a valley: a temperature inversion. Where I lived as a boy, on quiet winter mornings, I regularly hit a layer of air so cold that it made me gasp.

To feel the heat of air friction, cyclists would have to break a lot of records and a lot of bikes. Even at about 900 kilometres per hour, the skin of a cruising commercial jetliner only becomes about 30°C warmer than the air in the upper troposphere – still well below freezing.

At supersonic speeds, aircraft do get much hotter. Concorde routinely got too hot to touch: more than 120°C near the front and more than 80°C towards the back. In fact, its maximum cruising speed was limited not so much by its power as by the temperatures that its aluminium skin could tolerate. Such temperatures distorted its airframe, but not as badly as the Mach 3 Lockheed Blackbird, which reached well over 300°C, requiring special engineering and construction materials to cope with the expansion the plane underwent as it heated up during flight.

Because the Blackbird lacked sealants that could withstand operational temperatures, the plane actually leaked fuel until it warmed up after take-off. Fortunately, as a cyclist you will not feel frictional warming below speeds at which the slipstream would take your skin off.

Jon Richfield
Somerset West, South Africa

An object entering the atmosphere is not heated much by friction directly, but by the rapid compression of the air that it passes through. Friction – that is collisions – between the molecules in air generates heat, which passes to the object travelling through it. It is similar to your pump warming up as you inflate your bicycle tyres. The effect only becomes significant at supersonic speed, though, when the compressed air cannot get away.

Peter Mabey
Harlow, Essex, UK

Sticky ice

? I was watching the bobsleigh event at the Winter Olympics recently and it was foggy. The commentator said that extra humidity caused by the fog would 'stick to the track' and slow the competitors down. How? I would assume extra moisture in the air would make the track more slippery and thus faster.

Aremenis Pandyu
London, UK

The essence of success in a bobsleigh race is speed. Ice is laid down on the concrete course in millimetre layers over several days to a thickness of around 5 centimetres. The concrete is networked with plumbing, through which ammonia refrigerant flows, keeping the ice at about –6°C. The ice meister, who oversees the track, uses sensors in and around it to keep informed of the conditions, so all aspects of the course are controlled and optimised for speed.

Nature is harder to manage, and humidity can pose a problem. If the temperature drops towards the dew point, moisture condenses from the air and a crystalline frost forms on the ice. This hoar frost acts as a brake under the sleighs' runners. Before racing, the track is shaved and sprayed to ensure no frost or bumps remain to slow the sleighs.

You can experience frost's unlikely friction if you walk on

an icy path in the evening, then walk on it again the next morning with a fresh covering of hoar frost. You will find that grip is greatly improved.

David Muir
Science Department, Portobello High School, Edinburgh, UK

8 Planes, Trains and Automobiles

A bed of rocks

? Railway sleepers often sit on a bed of small stones that act as ballast. This ballast material stretches well beyond the width of the sleepers and to quite a depth. But why does it have to be stones? Could anything else be used either physically or economically?

Peter Brigg
Queensland, Australia

Ballast provides a firm surface for the railway track and spreads the load on the subsoil. To do this it also needs to drain well. Ballast that is of sufficient depth will distribute the load. And to restrict lateral movement of the rails, ballast extends outwards in shoulders on either side of the track.

The techniques of ballasting have developed over hundreds of years, starting with road building before it was used in railways. It is difficult to imagine a material that would do the job as well or as economically as stone ballast. A suitably hard stone will stand up to a lot of wear and tear. The random nature of the broken stone means that it tends to lock into a formation in a way that, say, spherical granules would not. And, of course, the spaces between the stones allow for free drainage.

Ballast is not without its problems, however. The stone has to be a suitable grade and it has to be laid with care and accuracy. Even so, it will degrade over time. It also settles and this must be tackled by adding more ballast and tamping (or packing) it into place. Drainage can be inhibited by dirt, including soil, contaminants from passing trains and even stone dust worn from the ballast itself.

At one time the ballast would have had to be replaced with new material, but now there are special trains that can remove the ballast, clean it and tamp it back in place. But ballast is not the only way of achieving a firm, level track. Track laid on concrete beds is increasingly common. With modern machinery this can be laid quickly and easily and is very durable. It is more expensive than ballasted track, but it has features that may make it desirable in special cases. One common example is in a tunnel, where the track has to be lowered to allow for overhead electrification. Because the concrete bed is shallower than ballast would be, it can be laid in place without digging out the base of the tunnel.

Glyn Williams
Derby, UK

Glyn Williams discusses ballast without referring to the nature of the rock used, which is a critical parameter. Ideally the rock should be from a tough and stable lithology. Samples of the potential material are examined in thin sections by a mineralogist using a petrographic microscope. They will be looking for evidence of minerals that are likely to break down when exposed to air, water, load pressure and so forth.

In some states in Australia, any ideal rock would have to

be transported over vast distances and much of the ancient rock available at the surface is weathered. This means that for economic reasons, the ballast used is often less than ideal. Mine sites are useful because mine waste brought to the surface is not weathered in the same way. However, this kind of material can cause other problems.

In one example, ballast taken from a nickel mine caused a serious degradation of metal sleepers, because it contained significant quantities of pyrrhotite, a very unstable form of iron sulphide. The pyrrhotite oxidised, producing sulphuric acid, which rapidly corroded the metal.

Concrete beds are not a foolproof solution either. The ubiquitous 'concrete cancer' – where the cement paste reacts with the rock aggregate, causing cracks and spalling – directly relates to the unsuitability of the rock used.

Roger Townend
Townend Mineralogy Laboratory, Malaga, Western Australia

Twisted tips

? Many aircraft wings have their tips turned up to reduce turbulence. I have not observed the same on wind-turbine blades – do they not experience the same problem?

Andrew R. Doble
Hamilton Parish, Bermuda

The upturned tips on aircraft wings are called winglets, and are intended to reduce drag, rather than turbulence. Wings generate lift through the difference between low and high pressure on the top and bottom surfaces, but at the wing tip, these two pressure fields meet, causing air to flow from high pressure to low. This flow generates a tip vortex, which drags airflow at the rear of the wing downwards. Lift acts perpendicular to this airflow, so if the flow is dragged downwards at the trailing edge, a component of the lift force now pulls backwards, opposing the forward motion of the wing. This is known as induced drag.

Wing shape can be designed to minimise induced drag, for example, the beautiful elliptical wings of the Second World War Spitfire fighter plane. Winglets, on the other hand, create a physical barrier to airflow between the upper and lower surfaces, reducing the effect at the cost of increased friction and mass at the wing tips.

The longer and thinner a wing is, the less the bulk of the wing is affected by what's going on at the tip. Because wind turbines tend to have very long, thin blades, the structural disadvantage of adding winglet mass at the tips outweighs any reduction in induced drag.

Daniel Summerbell
Cambridge, UK

An earlier answer about how winglets – the upturned tips on aircraft wings   work perpetuates an unfortunate misunderstanding about the way wings generate lift. It states, almost as an aside, that the pressure difference between the top and bottom of the wings causes lift.

There is indeed a pressure difference there, which answers the winglet question nicely – as the response explained very well. But the assumption that this difference also causes lift misses the mark.

The better and simpler explanation draws mainly on

Newtonian physics: in basic terms, a wing or aerofoil tilted slightly upwards at the front deflects air downwards. This flow produces an equal and opposite reaction: lift.

Paul Hargreaves, private pilot
Windsor, Ontario, Canada

Cancelled out

[?] In early November, UK airports cancelled many flights because of heavy fog. But why does fog still disrupt aircraft departures and arrivals? Surely we have the technology to prevent this.

Alan Penrose
Northampton, UK

During normal landings, pilots take over manual control of the aircraft at or above a 'decision height', which is typically 200 feet above the runway. When the cloud base is lower than this, or when fog reduces the surface visibility, modern airliners are indeed capable of landing themselves.

However, because air traffic controllers are unable to visually confirm that the preceding aircraft has vacated the runway, they increase the spacing between landing aircraft. Likewise, to prevent interference to the landing-system transmitters, controllers don't allow departing aircraft to queue up at the end of the runway, so the rate of take-offs is also affected.

This reduces the number of aircraft that can take off and land per hour. At very busy airports that are operating close to the limit of their runway capacity, the only way to achieve a lower take-off and landing rate is to cancel some flights and delay others.

For this reason, less-busy airports are less likely to have delays and cancellations during periods of poor visibility. Airports such as Heathrow and Gatwick, which plan some of the highest number of take-offs and landings per runway per hour in the world, are particularly hard hit.

Steve Broadbent
Carterton, Oxfordshire, UK

In 1972, I made my first flight in a large commercial aircraft – a VC10. As a glider pilot, I was interested to see how the pilot would make his approach and landing at Dulles Airport near Washington, D.C.

I had a window seat, it was a clear day and I watched as we came in with a long, low, sweeping approach and finally came to a stop at the far end of the runway. 'Curious,' I thought. The pilot then announced: 'Ladies and gentlemen, I thought you would like to know that this aircraft is fitted with equipment that enables it to land by itself in fog, so I thought I would give it a go. It failed!'

Brian Cordon
Matlock, Derbyshire, UK

Off the rails

? I was looking out of a train window at the live rail on the adjacent track during a heavy rainstorm. This made me wonder how much electrical power is lost across the wet ceramic supports. Is it enough to cause a significant voltage drop in the locality where the rain is falling?

Jeff Blyth
Brighton, UK

My experience is with electricity supply, but the principles involved in the transmission of high-voltage electricity in heavy rain are similar to those in electrified rail systems.

Rather than speaking about power being 'lost', it makes more sense to talk about the electrical current leaking across the ceramic supports or the insulators. It should be noted that, contrary to the impression one might get from television, rainwater is a fairly poor conductor of electricity. The track insulators are also designed in a special shape so there is no continuous wet surface between the live conductor and earth. These factors make the leakage current small compared with, say, the current drawn by a train starting from rest.

Peter Smith
Long Ashton, Somerset, UK

Tired tyres

? Standing next to a busy highway most of the noise comes from the car tyres. Is there any way to reduce this noise while keeping the more desirable properties of tyres such as traction?

Don Jewett
Mill Valley, California, US

There has been a great deal of research into this topic in the EU over the past twenty years. During the time I worked on traffic noise at the UK's Highways Agency, a range of quieter surfaces was developed. Long-term trials were undertaken to monitor their resistance to wear and skidding performance as well as the amount of noise emitted in a standard roadside test.

Part of the problem arises as tyres interact with surfaces

that are deliberately roughened to improve skidding resistance. Concrete has been used extensively in the United States because it is durable and economical for long stretches of road, but it is difficult to create smooth rolling surfaces in concrete while ensuring the right texture for good skidding resistance.

Efforts have been made in the US to retexture some older concrete roads using diamond saws to cut longitudinal slots in the surface that help to disperse water. The result is that tyre noise is reduced and may be less than after the transverse-brushing process (to provide skid resistance) used on concrete highways in the UK. However, sawing and regrinding concrete is a slow and expensive process.

Peter Kinsey
Llangybi, Monmouthshire, UK

The noise comes from tyres' interaction with the road surface. Hence the nature of the surface can make a big difference to the noise produced. Many motorways in my native country, the Netherlands, are being surfaced with ZOAB, the Dutch acronym for very open asphalt concrete. The surface's high porosity allows sound to dissipate downwards as well as upwards and sideways, creating markedly less noise. The dampening effect is noticeable both outside and inside vehicles.

Another advantage is a reduction in spray in wet conditions. Traction is deemed to be acceptable on these surfaces but is reduced compared with typical road surfaces, especially when the ZOAB is new. That's why you will see signs saying 'new road surface; increased braking distance' along recently surfaced stretches of these motorways. The other downside is the increased need for salting in winter. The wet nooks and crannies in the ZOAB surface are perfect hidey-holes for ice. These road surfaces break up more easily as well, and need to be relaid more frequently than solid asphalt.

Jan Meulendijk
Haverfordwest, Pembrokeshire, UK

Feeling the pressure

? When we took a flight recently we noticed an unopened packet of crisps inflate because the cabin air pressure fell with altitude. At the same time, a member of my family commented that they felt bloated and uncomfortable. Is it possible that a similar effect was taking place in their gut?

Duncan Guthrie
Edinburgh, UK

The cabin air pressure at 10,000 metres – the cruising altitude of modern airliners – is about two-thirds of normal sea-level pressure. The contents of any gas-filled elastic container will expand due to this difference in pressure. Anybody who has flown with a heavy cold or with blocked sinuses may have experienced this effect, which can cause severe pain on ascending and descending as gases in the head expand. To answer the question about gut contents, most of the gas in the gut is in the large bowel and if it expands, there is only one way for it to come out. Some find the liberation of colonic miasma one of the surprising pleasures of jet travel. Perhaps on learning this readers will never view fellow travellers with equanimity. Of course, this would almost certainly be hypocritical.

Philip Welsby
Edinburgh, UK

Flying blind

? Do large modern jet airliners need windows for the pilots? In theory could a Boeing 747 be flown from London to Sydney without the pilots needing to see outside?

Peter Marston
London, UK

As civil aircraft are designed and certified at present, yes, forward-view windows are required. All take-offs are done visually and manually, keeping the aircraft straight on the runway by looking out of the forward windows to maintain track and most importantly to detect and correct the lateral swing caused by any engine failure. This in general terms requires a minimum horizontal visibility of 125 metres, which isn't much when you are travelling at 260 kilometres per hour.

Landings are different. Most modern airliners are capable of automatic landings, given suitable ground-based guidance. The minimum visibility for an automatic landing is generally 75 metres. This distance isn't for pilot visibility, but for safe access by ground vehicles to any incident. There is no minimum requirement for a pilot's view of the runway during such low-visibility automatic landings.

The bit of the flight between take-off and landing has no need for any external vision and it can be – and sometimes is – conducted solely on instruments. However, seeing and avoiding meteorological phenomena such as thunderstorms and their associated turbulence, hail and lightning is often best done by a combination of radar and visual scanning. All that said, synthetic vision systems are becoming relatively common and may well lead to aircraft where all external views are on screens and there are no windows. But please don't take away pilots' views, we are blessed with the best imaginable. I can spend hours watching the world go by beneath me. It's glorious.

Steve Moody, Airbus 320 Captain
Leamington Spa, Warwickshire, UK

The short answer is that pilots don't need windows to fly a plane. But it isn't the flying bit that we use windows for. While the climb, cruise, descent and sometimes actual landing can be done 'blind' using complex air and ground-based monitoring equipment, all take-offs – especially the part where the aircraft is in contact with the runway – are done visually, as is all taxiing around the airfield. The vast majority of landings are also done visually.

So-called blind landings, where the forward visibility at touchdown can be as low as 75 metres, require special ground equipment. They also involve vital operational procedures to ensure objects such as ground vehicles remain clear of transmitting aerials. All this markedly restricts the rate of aircraft movements on the ground, so are only applied when really needed. This explains why in fog, major airports such as London Heathrow restrict the number of flights they can handle, leading to cancellations.

Also, while all modern commercial aircraft have blind-landing capability built in, many airports don't, and as explained above even those that do only use it when necessary. However, as the questioner implies, once well away from the ground there is frequently little actual need to see out of the front window. Weather radar can show us thunderstorm cells,

allowing us to deviate around them and at the altitudes airliners fly there is generally no cloud thick enough to obscure
mountains (what pilots call cumulogranite), nor indeed mountains to bump into.

I have never flown Concorde, but I believe that once the
drooped nose (an engineering necessity to allow forward
visibility during landing) was raised to the cruise position,
a shield slid up over the forward windows. This improved
the aerodynamics but obscured the view. However, even in
this highly technical age, there will always be a need for
visual flying, and it would be a shame to lose some of the
best views on earth.

Martin Powell, Captain, Monarch Airlines
Westgate-on-Sea, Kent, UK

Airliners do need windows, there is no autopilot function
that would taxi the aircraft from its stand to the runway and
then take the aircraft off. Guiding an aeroplane through
take-off is a visually demanding task and the pilot flying the
aircraft initially has their eyes 'glued' to tracking the runway
centre line. In principle, we can see how this could one day
be automated, but there is more to the procedure than simply
providing guidance to the aircraft's automatic systems. For
example, flocks of birds or lost light aircraft might be seen
and avoided during the take-off. Indeed, as in the case of US
Airways Flight 1549 that ditched in the Hudson river close
to passing rescue craft, I think we are a very long way from
developing automatic systems capable of producing such a
flexible response.

Shortly after take-off, the autopilot is typically engaged
and so, again, in principle and all being well (which it often

isn't), most of the route could be flown without visual reference. However, today most airliner flight paths are not flown as planned. During the flight, air traffic control can pass a number of route changes to the aircraft that may require a view of the outside world, the pilots may also change routing on the basis of something they have seen, such as other aircraft, volcanic ash clouds, etc.

It is important to appreciate that airliners do not always fly in controlled air space, and so avoidance of collision with other aircraft is at times dependent on pilot vision. During all phases of flying, there is the possibility of failure of the automatic systems when their control parameters are exceeded. In these cases the pilot needs to take over the control function, and this can require a view of the outside world.

The effects of extreme space weather events such as massive solar flares have been in the news recently, and potentially there could be a number of effects on the aviation industry across areas such as power distribution, radar, telecommunications, GPS, computing, flight control, health effects, and so on. It is perhaps more likely that power supplies to ground infrastructure electronic dependent navigation and telecommunications may be affected. On this basis, provided flight control is maintained you will still have pilots who know the time, know where they are, have a map of where the airports are, and who have been taught old-fashioned navigation techniques, so provided they have a window they can see out of, the situation is not so bad.

Rob Hunter, Head of Flight Safety, British Airline Pilots' Association, via the web

Spinning wheel

? On a London Underground platform recently I saw a man wearing an official badge holding what looked like a small spinning fan. I realised that it was powered by the wind coming through the tunnels ahead of the trains. He was obviously measuring wind speed. Why was he doing this?

Justin Davies
London, UK

The spinning fan is an anemometer, and the man was measuring the ventilation provided by the passage of trains. This plays a great part in keeping the air in the Tube tunnels fresh, although other means of ventilation are also used. I inherited the same type of instrument from my late uncle. He was a mine safety officer and used it to measure the ventilation in a coal mine. Curiously it does not measure wind velocity directly, but instead measures a 'length of wind'. The device is graduated in feet and is timed over a minute to give the length of the air column passing through in that time. The lever at the top engages the clutch to start the measurement once the fan has got up to speed.

Tony Finn
Hull, East Yorkshire, UK

Railing against it?

? Edinburgh finally got its tram system last year which was over-budget and behind schedule. Most of the problems were due to having to lay the rails in the

road. Why didn't Edinburgh and other UK cities go for
trolley buses – electric buses powered from overhead
wires – instead of trams?

Jim Logan
Castle Douglas, Dumfries and Galloway, UK

From my experience in having delivered a rail-based road-
running urban tramway within budget, I can think of several
reasons why trolley buses were not preferred. Unguided
trolley busways with the same passenger capacity as tramways require far more space. Drivers of trolley buses cannot
match the precision of a vehicle guided by rails, so they need
more space as a safety buffer. Guidance systems for buses
that don't run on rails do exist, but their use of space is not
as efficient as that of a tram. Such guidance systems may
also prevent other traffic from using the busway and so cause
congestion.

The two systems suit different circumstances. For example,
a tramway might be the right solution for transporting high
numbers of passengers through limited space. As far as budg-
eting goes, few tramways or guided trolley busways are built
each year, so there is little experience to draw on when
identifying key cost issues. The detailed investigation needed
to establish the cost of a system is in itself very expensive
and potentially disruptive. Sometimes it might be practical
to carry out the exploration and construction at the same
time, and to accept the uncertainties in cost. But obviously
that decision is a political rather than an engineering one.

J. H. M. Russell
Golant, Cornwall, UK

A mast-see

? Tall masts and towers have warning lights to make them visible to aircraft pilots. But how does the autopilot system know of the presence of such tall, thin structures? I assume that the autopilot cannot see the lights.

Perry Bebbington
Nottinghamshire, UK

The simple answer is that the autopilot cannot see the tall masts. So why do planes rarely hit them? The reason lies in the fact that there are two types of flying protocol: visual flying rules (VFR) and instrument flying rules (IFR).

In VFR flying it is the pilot's responsibility to see and avoid obstacles, so it is only allowed when visibility is good. Commercial airlines must fly under all conditions, so they always fly under IFR. All aircraft using those rules reserve a flying pathway through the sky: these permitted routes are known to be completely free of all barriers such as tall masts and mountains. The autopilot serves only to keep the aircraft safely flying along its pathway. Controllers on the ground by means of radio communications to the pilot manage the separation of IFR aircraft by allowing only one plane at a time into an individual path segment. The system works remarkably well.

It is a legal requirement to keep new towers from intruding into IFR routes. If a tower must be built in one, then it can only be constructed after the pathway has been adjusted to avoid the proposed tower. So tower avoidance in IFR flying is a matter of planning the flight to stay in the permitted airspace. Should the pilot ask the autopilot to fly outside that safe airspace, there is nothing at all to prevent the plane from hitting a tower or even a mountain.

A further problem is when a rogue plane ignores the rules and intrudes into the reserved airspace. Hopefully in such situations the controller on the ground will see this on their radar and tell the IFR pilot to immediately change course and avoid a possible collision.

Ed Gauss, former bush pilot
Niwot, Colorado, US

Blowin' in the wind

After videoing something outside, I found that although I hadn't noticed it being windy at the time, the recording played back a very obvious wind noise. Why did the wind noises seem much louder during playback?

Hannah Edwards
Sheffield, UK

Wind noise has a spectrum that is tilted massively towards low frequencies. By contrast our ears' sensitivity drops off steeply in the lower octaves. Thus we tend not to hear wind noise as very loud in normal life. As a designer of microphone windshields I know that mics – particularly modern ones – can have a frequency response that is nearly flat even at very low frequencies. If they are used without an adequate windshield, the low-frequency content of wind noise can easily drive them and other audio equipment into overload, even though it doesn't sound very loud to our ears.

Overloads at frequencies too low for us to hear lead to short-term gaps in the recording during which the amplifier suppresses all incoming sound. The result is a sort of syllabic muting effect or whumping. This is far more intrusive than the simple sound of onrushing air and is the characteristic recorded 'wind noise' that your correspondent objects to.

Chris Woolf
West Killatown, Cornwall, UK

In my spare time I work as a sound engineer for a group of amateur video-makers. We find that wind noise can be a problem when recording on location. The nature of the sound heard will depend on the shape of your ears, and the direction in which you are facing relative to the wind. Conditions will be different for the microphones fitted to the camera, causing the sound heard on playback to vary from what your ears pick up. In addition, the brain is accustomed to processing background noises and will suppress them to the point that they are almost unnoticeable.

When your brain processes the sounds on the recording, however, they are suddenly very apparent. Sound engineers cover outdoor microphones with a furry 'wind-gag'. These are sometimes referred to as 'dougals' because they resemble the shaggy dog of that name on the children's TV show *The Magic Roundabout*. They have the effect of reducing air velocity while allowing the pressure changes that carry the sound to pass through. On location, a sound engineer will always monitor the recording via headphones and so will be more aware than the rest of the crew of unwanted sounds such as wind noise and the rumble of passing vehicles.

Chris Finn
Beverley, East Yorkshire, UK

Air drumming

? When I open my car windows while driving, everyone inside can hear a helicopter-like thumping sound. It is even worse when the rear windows are opened and it gets louder at higher speeds. What causes this phenomenon?

Oscar Holroyd
Herefordshire, UK

An open sunroof or window essentially acts as an acoustics phenomenon known as the Helmholtz resonator. The most common example of this is the sound created when you blow across the top of an empty bottle.

The pressure inside the car resonates according to the cross-section of the opening, the speed of the air over it, and the volume of the cavity inside. These changes in pressure are perceived as a drumming sound. Imagine that the air inside and outside the car is separated by a thin invisible membrane. If something were to push down on the membrane and let go, the membrane would spring up and down with the amplitude of the oscillations getting smaller as with an elastic band.

But what if a downward push arrived just as the membrane was already moving downwards? This is where vortices come into play. They create resonance, maximising the amplitude by pushing down and increasing the pressure in the car just as the imaginary membrane is moving downwards anyway. This is similar to the way that people time their push on a child's swing. You push hard just as the swing is starting to move downwards again to get maximum thrust. The resonant frequency is shifted higher for bigger openings and smaller volumes, and vice versa. So blowing across a half-empty bottle should generate a higher frequency than blowing across an empty one.

Mike Follows
Sutton Coldfield, West Midlands, UK

Treading lightly

? By UK law car tyres must have a tread at least 1.6 millimetres deep. On my new tyres it is about 9 millimetres. At what point would deeper treads worsen roadholding and why?

Jack Shaw
Barnsley, South Yorkshire, UK

The maximum tyre tread before the ride becomes too rough depends on the kind of road you're driving on and what suspension you have. Bicycle tyres are a good example – a thick knobbly tyre grips mud and loose terrain well, but is heavy and noisy on tarmac. The heavy tread means that it won't rebound so quickly to the relatively small stones on the road, so its grip is poor.

Modern tyres on racing bicycles are thin and light. The tread layer has low mass so it can follow small road undulations providing good grip. Because the internal pressure of the tyre is high, it pushes through any water so even slick tyres grip well in the wet. On a good road you can do over 40 miles per hour in comfort. But the tyres have zero grip on slimy mud when cycling.

Simon Dales
Oxford, UK

Tyres do not need grooves to improve grip in dry weather. This is demonstrated in motor racing in which smooth 'slick' tyres are the rule unless it is wet, when the addition of grooves allows a tyre to squeeze away water that would otherwise lubricate it and remove grip. The heavier the rain, the deeper and more numerous the grooves need to be.

New road tyres have sufficient grooves for all but very heavy rain. But as they wear, their capacity to shift water drops because the grooves become shallower, until they are insufficient even for light rain – which is the limit for legal tyres as noted in the question (with a safety margin added).

The deeper the grooves, the less grip the tyres have because less rubber touches the road and the rubber between the grooves moves about. So on a dry road, any groove inhibits grip. But unless you can do a pit stop as soon as it rains, you are still better off with grooved tyres for road use, purely to ensure you can drive in all conditions.

John Davies
Haverbreaks, Lancashire, UK

Upside drone

? Why do military drones look so different from military aircraft? For example, their tail fins point downwards.

Martin McCann
Worcester, UK

Abraham Karem was the designer of the Albatross prototype, which later became the Predator drone. This is probably the drone we are most familiar with from news reports on TV.

He notes that drone makers have a distinct advantage in not needing to consider accommodation or life support for a pilot.

Karem designed for endurance, giving his drone long wings to maximise lift and stability over speed and agility. He chose a propeller-driven engine because these are more fuel-efficient than jet engines, and usually easier to maintain. He located the propulsion at the rear of his drone to minimise any interference to sensors housed in the nose. The distinctive split downward-facing tail of the Predator exists to protect the propeller during take-off and landing. If the pilot attempts to come in at too steep an angle, the outer tails act like a pair of skids to prevent the propeller from striking the ground. The nose is bulbous to accommodate the satellite-communications equipment required to remotely pilot the drone.

The Predator was not designed as an armed aircraft, but for use as a surveillance system. There are many other designs for drones out there. Not all use propeller propulsion, and some are designed to benefit from modern stealth technologies. There is also work going on to develop un-crewed, remotely piloted combat aircraft. These will probably look more like crewed versions because they will need to be far more agile.

Thomas Woods
Eyemouth, Berwickshire, UK

Aerial drone

[?] Why do light aircraft with piston engines make so much noise, even when they are far away? They can be very intrusive, especially in secluded landscapes such as Dartmoor National Park. Are they fitted with silencers like road vehicles? And if not, why not?

Duncan Hutchinson
Newton Abbot, Devon, UK

Designers traditionally gave little thought to noise in small aircraft, concentrating instead on lightness, simplicity, payload and performance. Research has shown that the engines and propellers make most of the noise. Silencing them for military reconnaissance proved very effective during the Vietnam War, for example – the Lockheed YO-3 was practically inaudible at an altitude of 300 metres. However, muffling the engine, especially the exhaust and air intake, remained costly in terms of weight and performance: the efficiency of internal combustion engines is sensitive to how well they breathe.

Silencing engines proves less demanding than silencing propellers, for which blade width, length, shape, rotation rate, choice of material (wood, metal or synthetics) and number of blades all proved relevant factors to consider. Each of these also affects cost, weight, performance and safety. In the competitive business of small aircraft manufacture, noise abatement and sophistication understandably gets lower priority than cost and proven design.

However, as the numbers of airfields and light aircraft in population centres increases, the demand for sound abatement is intensifying, so muffled engines and cheap, efficient, scimitar-shaped blades can be expected in new standard designs.

Jon Richfield
Somerset West, South Africa

I can suggest a number of reasons why light aircraft make so much noise. The first is that they are predominantly flown as a hobby, so cost is a major factor. Therefore light aircraft need to be just that: light. The heavier a small plane is, the bigger the wings and engine need to be, all other things being equal, and the more expensive it is to run.

Adding silencers – most don't have them – incurs undesirable cost, as well as complexity and weight. Also, they reduce engine power, and since light aircraft tend to be built with small engines anything that reduces power is highly unwelcome.

Finally, small planes tend to fly slowly, relatively close to the ground and below the clouds. This means the noise is louder at ground level than it would be if the plane were higher and above a muffling cloud layer.

Manek Dubash
Lewes, East Sussex, UK

Ring of confusion

? On 22 June this year in north Pembrokeshire, UK, I spotted a thin circular cloud in the sky, thicker at one point. There were aircraft trails around – but all of them were absolutely straight. So what is this?

Bob Holmes, via email

This is undoubtedly an aircraft condensation trail (contrail) made quite deliberately. The dense knot is where the aircraft was climbing (or descending) almost vertically, concentrating the contrail in one spot. The aircraft levelled off then turned abruptly before starting the circle. Once it had arrived back at the knot it closed the loop by descending (or climbing)

206 How Long is Now?

steeply. The transition between vertical and horizontal circling is evident in the convex shape of the trail at this point.

An old friend of mine was a navigator in two-seater fighter aircraft. He told me how on a day when conditions were ideal for contrails to persist he navigated his pilot through a complex series of manoeuvres and produced a good representation of the male anatomy. His aerial graffito was apparently visible (but not necessarily appreciated) over a large area of southern England.

Jack Harrison
Tobermory, Isle of Mull, UK

9 Technology

Fluff stuff

?   Why do some woollen cardigans or jerseys produce a lot of pilling whereas others don't? Is there any way to know beforehand if a wool garment will be prone to pilling?

Gabi Simon, via email

Pilling is caused by short fibres working themselves to the surface of the fabric. There, friction from whatever the fabric is in contact with causes the scales on each fibre to lock to those on other fibres, with the result that they become matted into a ball. There are several contributing factors.

The first is the nature of the fleece. The fleece of some sheep breeds such as Southdown has short fibres, typically between 4 and 6 centimetres long, while long-wool fleece such as Wensleydale can be up to 30 centimetres. It is easier for a short fibre to work its way to the surface of a fabric than a long one. And some fleeces are easier to felt than others; Ryeland with a staple length of 5 to 6 centimetres is famously difficult to felt, while merino, which has a similar length, felts readily.

Loose fibres that will not felt cannot form pills. A second factor is the way that the fleece is prepared and spun into wool. Hand spinners typically prepare short-fibred fleece by carding it into rolls of fibre called rolags, then spinning it in a 'woollen' style. This causes the fibres to become randomly arranged and the resulting yarn is light and full of air, making it easier for fibres to work to the yarn's surface. Long-fibred fleece is typically prepared by combing, which removes the shorter fibres. It is then spun in a 'worsted' style which keeps the long fibres parallel and smoothed down, thus making it hard for them to escape the yarn and cause pilling.

The amount of twist in a yarn also affects its tendency to pill; a higher twist holds the fibres more firmly and prevents them escaping. Many commercial knitting yarns are quite lightly spun, with little twist, whereas a yarn intended for weaving – often spun from long-wool fleece – will usually have a much higher twist.

Finally, the way the yarn is turned into fabric can affect pilling. A dense fabric will hold the fibres firmly and have less tendency to pill, while a loose one will allow them to escape. Seeing as the questioner mentioned woollen fabrics, I have described features of sheep's wool. Other fibres, for example alpaca, or synthetic fibres such as acrylic, are not officially called wool but the same principles will apply. To see if a wool garment will be prone to pilling, examine the fabric to see how dense it is and how tightly the yarn is twisted – the looser it is, the more likely it is to pill.

Judith Edwards, hand spinner
Cumbria, UK

Heavy metal

? Does a new MP3 player weigh less when it has no music on it?

Paula Maxwell
London, UK

The music that is stored on an MP3 player is represented by a particular sequence of 1s and 0s. When music is put onto the device, this data is kept in the player's memory.

In most MP3 players, the memory is a solid state chip made up of an array of transistors. As the electrical current flows across the array, a value of 1 is encoded by the presence of electrons in a transistor, while a value of 0 is encoded by the lack of electrons.

We might expect a brand-new solid state memory device to start out being filled with 0s, and that encoding any meaningful data would involve adding electrons. This would then make a new player with music files on it weigh more than one without music, because electrons have mass. In fact, new memory devices typically start out full of 1s, so it would be easy to assume that players are heavier before songs are added.

But because the current comes from the player's battery, which forms a closed system with the rest of the player, there is no overall gain or loss of electrons – and the weight remains unchanged. Electrons are simply redistributed.

Even if individual MP3 players were open systems, a group of players en masse would not gain weight, because some files will have more 1s than 0s while others will have more 0s than 1s – so their masses should even out.

But such devices will measurably – if impalpably – gain weight as their users put music onto them, because of the sweaty or greasy deposits left as they are handled and the dust and debris that sticks to these deposits. Alternatively, rough treatment can scratch away surface lacquer or crack bits off, and this kind of chip would make the device lighter.

So, while neither heavy metal nor light choral will affect its recorded weight, how you treat your MP3 player through the day could certainly impact on its evening mass.

Len Winokur
Leeds, UK

An earlier reply regarding the weight of an MP3 player is misleading, although I agree with the conclusion.

It implies that data bits are stored by adding or removing electrons from a memory cell, and that these electrons are supplied by a battery. In fact, bits are stored by moving electrons from one part of the memory cell to another, so there is no net gain or loss of electrons in the memory chip. The chip contains the same number of electrons, and thus weighs the same, whether it's all 0s, all 1s or full of data.

Likewise, when a battery supplies an electron from one pole it consumes one at the other. When the battery is flat it contains just as many electrons as when it was fully charged. No electronic devices accumulate electrons; the flow in to the device is always balanced by flow out.

Doug Fenna
Ryde, Isle of Wight, UK

Aggressive pitch

? Why does feedback between a microphone and speaker generate a high-pitched noise, and not a low one?

Tony Sharpe
Almonte, Ontario, Canada

Sounds can be viewed as combinations of sine waves of different frequencies. The Barkhausen stability criterion states that any frequencies that perfectly fit in the system's 'distance' (from microphone to speaker and back) and are amplified along the way will be sustained and amplified further.

This distance is hard to pin down because it depends on factors like delays in the electronics, the room's acoustic properties, the positions of the microphones and the resonant frequencies of the instruments and speakers. But higher frequencies are more likely to enter a feedback loop because the waves are shorter. The odds are higher that you can perfectly fit many short waves into a certain distance than many long waves. So although you do get throbbing low-frequency feedback, high-frequency feedback is more likely.

Moving a microphone near a speaker causes feedback because live speakers are never truly silent: close up, you can hear a multi-frequency hum. And as you move the microphone, you're changing the distance and sweeping the range of frequencies that cause feedback, resulting in a high-pitched whine.

Ron Dippold
San Diego, California, US

As a musician, I have been a victim of feedback at all points on the frequency spectrum. There are two main reasons why higher frequencies resonate more often. First, most microphones, amplifiers and loudspeakers are designed to be less responsive to lower frequencies than the higher ones of voices and guitars.

Second, lower frequencies need more space to become 'in phase' and for their waveforms to flow back and forth. In smaller venues, there is not enough distance between the microphone and speakers for this to happen. Once a stage is large enough, it is more common for bass frequencies to resonate (and for band members to complain about them).

Connor Fitzgerald
Berlin, Germany

Sloping off

? On my train to work there is a dot-matrix sign which has the words 'Calling at' in a Roman font, but the station names below it in italic font. However, if you take a photo of the sign, all the words appear Roman. If the photograph catches the sign at the correct moment, you can see that the italic font is created by the dots on the sign lighting up with some clever on–off timing that fools the eye. All the dots are in vertical and horizontal rows. How is this italic effect created and, more intriguingly, why bother?

Craig Mackie
Easton on the Hill, Northamptonshire, UK

The italic effect that the questioner describes is not created deliberately. It is an accidental result of the way the dot-matrix display works, through a technique called multiplexing. The number of LEDs in the display is large – easily in the thousands – and it would be expensive to use a separate electronic driver to control each one. Instead, the lights are controlled by column and row. Each LED is illuminated by powering its column and row drivers at the same time – a bit like the game Battleships. And although the display appears to be illuminated steadily, it is actually flickering at a high speed, with only one row lit up at a time.

This is fine until it becomes necessary to show a moving line of text. To avoid the words distorting – in this case the display would have to flash its entire sequence of rows, perhaps twenty-five of them, before the letters changed position. By using this method the text would appear to move only very slowly. If the text is to speed up, the display only has time to illuminate a few rows (or maybe even just one) before the position of the letters changes. When the subsequent row flashes up, the text has moved a little to the left or right, and this leads to the apparent sloping lines.

The alternative is to refresh the display one column of LEDs at a time, but this would make the distortion even worse. Instead of an acceptable italic slant, the characters would appear to be squashed or expanded and would not be legible. The reservation displays above seats in some trains demonstrate the problem, with the text seeming to move infuriatingly slowly. There could be several reasons for this, but one may be that the displays are driven a column instead of a row at a time, meaning that fast-moving text cannot be accommodated at all.

David Gibson
Leeds, West Yorkshire, UK

Windblown

> **?** Our local community turbine rotates at a constant speed of 39 revolutions per minute. What is the mechanism that allows greater power to be generated in strong winds, in terms I can explain to my grandchildren?

Glyn Hicks
Fishguard, Pembrokeshire, UK

For good reasons, our electricity supply works not by providing a single, steady flow, but by flitting backwards and forwards. This is called alternating current, or AC for short. The electricity generated by wind turbines must keep in time with this, so all of them rotate at the same speed. When the wind blows harder, adjustments inside the turbine make it harder to turn the rotors, generating more electricity with each rotation.

It is exactly like trying to pedal your bike at a continuous speed. When you leave the flat and start going uphill, it becomes much harder to maintain the same pedalling rate.

Tim Stevenson
Great Missenden, Buckinghamshire, UK

Older designs for wind turbines aim to keep the blades rotating at a constant speed because this enables optimal conversion of the movement into electrical energy for use in the national grid. Too slow and the turbine doesn't catch enough wind energy; too fast and the blades and other components suffer excessive stress. If left to itself, the turbine would spin faster as the wind increases, just like the plastic windmills we buy at the seaside.

We convert the wind into electricity by allowing the

turbine to spin freely and then 'braking' it to capture as much energy as possible. But instead of a mechanical brake such as those on a car or bicycle, which remove the energy from a spinning wheel, we use an electrical 'brake' that allows us to capture energy and convert it into electricity rather than waste it as heat. We don't want to stop the turbine completely, but allow it to turn at a constant speed regardless of how hard the wind is blowing.

If we want to freewheel a bicycle down a gentle hill at a steady speed, we apply the brakes gently and they warm up by the time we reach the bottom. If the hill becomes steeper, but we want to keep the same speed, we squeeze the brakes harder and they get even hotter because they convert more energy into heat (don't test this by touching the brakes).

Similarly, when we want our turbine to turn at 39 revolutions per minute in a steady breeze, we only apply a small amount of electrical braking, and generate a small amount of power as a result. When it's blowing a gale, we can really apply the 'brakes' hard to get a lot more power out, even though the blades are spinning at the same rate.

Steve Agar
Hexham, Northumberland, UK

Scaredy cats

? How do ultrasonic devices built to scare away animals such as cats work? Do they mimic a high-pitched sound known to scare such animals, or do they scare them simply by their loudness? Perhaps it is just that the sound annoys the animals as much as it annoys me.

Lewis O'Shaughnessy
Salisbury, Wiltshire, UK

The ultrasonic cat repellers mentioned in the question sound similar to devices sold in Australia to repel rats, spiders, and crawling insects. I've found these devices work very well. The salesman said they emit a sound that is annoying but not harmful to the pests. He also recommended waiting up to ten days for them to move out. After installing some in our office, all the spiders had moved out of the toilet cubicles by day four.

Occasional visiting spiders remained for less than a day. We saw no fresh rat droppings either. Years later there was an unidentifiable annoying screech in one workroom. It occurred at random intervals audible for ten seconds every couple of minutes. I had hoped to work there but was driven out after forty minutes. After four days I couldn't enter the room at all. I had been conditioned to associate the sound with discomfort. We eventually found the faulty repelling device behind a bookcase. For those who wish to repel human pests, the screech frequency was somewhere between 4 and 6 kilohertz.

Peter Torney
St Kilda, Victoria, Australia

These devices offer a modest deterrent effect – at best. A cat's personality and whatever it sees as its incentive for straying into a garden are likely to trump most attempts to keep it away. A trial in suburban gardens suggests that these gadgets can cut the frequency of cat intrusions by about a fifth, and that those that do occur last a third less time. The combined effect is that cat traffic is halved.

Another trial, this time in the lab, introduced cats to the test area well beyond the device's stated range with food arranged at 1-metre intervals from the device. Cats within earshot of it were less inclined to forage actively, but lingered longer than those at a distance. The only apparent sign of discomfort was an increase in ear-twitching.

Similar high-frequency devices built to deter loitering teenagers work on the premise that adults will be less affected as they may have lost the ability to hear high-pitched sounds. Other products play classical music or whatever genre the makers deem unlikely to appeal to the average teenager. There is an urban myth that what has become known as the 'brown note' can be used to disperse crowds. The idea is that people lose control of their bowels when exposed to sound at a frequency of around 5 to 10 hertz, too low to hear but supposedly able to resonate in the body.

This myth seems to stem from research carried out as part of the US space programme. There were concerns about the stresses astronauts would experience at launch so their physiological responses were tracked while they were strapped into their cockpit seats and subjected to mechanical vibration. The astronauts experienced serious bowel and other bodily discomfort because the vibration could reach their bodies via the seats. Air, in contrast, is poor at transferring low-frequency vibration as sound.

In cowboy films characters may put an ear to railway tracks to 'listen' for an approaching train supposedly audible through vibrations travelling through the rails. In fact, Hollywood is guilty of poetic licence, because the vibrations can only be felt, not heard. Needless to say don't ever try this yourself, whether the railway is electrified or not.

Mike Follows
Sutton Coldfield, West Midlands, UK

Cloud cover

? How much data does the World Wide Web hold?

Colin Singleton
Sheffield, UK

Whatever the amount is in any one moment, it'll be quite a bit more in the next. A good estimate floating around is 1 yottabyte or a trillion terabytes. Yet this figure dates from 2014, and because the web is widely described as growing exponentially, it is reasonable to assume it could now have doubled to 2 or more yottabytes.

We should also consider the so-called deep web, which encompasses anything that is not found by a mainstream search engine. This includes many large databases for travel bookings, merchandise data for online shopping, any social media networks that do not put everything in the public domain, and so forth. Figures suggest that 80 per cent of the web is dark so if that is true the web comprised 5 yottabytes of data in 2014 and could now hold double that or even more.

It's also important to add that many websites do not reveal how much data they store. Hosting sites hold several redundant copies of almost everything, in multiple locations on varying types of media. Should we count this as part of the web? Similarly, much of the information on the web is duplicated. Years ago I searched for a list of jokes and found a dozen copies of the same list, with only minor differences, because every revised version was put on a new site and page. Who can guess how many copies of popular content exist? It's big, that's for sure.

David Morton
Geeveston, Tasmania, Australia

Radio gaga

? A car radio needs an external aerial to stop the car itself interfering with the signal. Yet the fob on the car key also works by radio, and is effective even if I hold it near the steel garage door with the car on the other side. So how does it work?

Adrian Somerfield
Dublin, Ireland

Metal is effectively opaque to all but the most energetic electromagnetic radiation, so most vehicles act as a Faraday cage, whereby the radiation is distributed around the exterior. However, some electromagnetic waves can pass through car windows – most notably visible light.

Remote keyless systems operate at a frequency of 433.92 megahertz in cars manufactured in Europe. This corresponds to a wavelength of about 0.7 metres, which means that the signal from a fob can pass through the windows to reach the transceiver inside. But some wavelengths are simply too big to travel through the windows, and for these the car might as well be a solid metal box. FM radio signals can be as short as about 3 metres, but AM signals range from more than 100 metres to half a kilometre. This is why cars are fitted with external aerials. Ideally their length should be one-quarter of the signal's wavelength, but this is clearly impractical for many wavelengths.

Mike Follows
Sutton Coldfield, West Midlands, UK

Ears tuned

? Hearing aids pick up airborne acoustic signals and can be switched to receive signals from an induction loop. Why haven't they been designed to receive TV and radio signals directly?

Adrian Smith
Headingley, West Yorkshire, UK

The main reason hearing aids don't pick up TV or radio directly is size. A TV aerial needs to be big to capture the signal, and it won't fit into a hearing aid. Even with a strong signal, hearing aids use very small zinc-oxide batteries which have limited capacity, restricting their ability to perform the necessary tasks.

However, the industry is working to solve these problems. Hearing-aid and technology companies are working with the Bluetooth standards group to transmit and receive audio over the low-energy version of Bluetooth. Your hearing aid won't receive TV or radio signals directly; rather the audio will be wirelessly transmitted from your TV tablet or phone to your hearing aid. The technology is also being designed to replace the current telecoil inductive loops in public places, providing higher quality along with the ability to listen to sound in stereo.

Nick Hunn
London, UK

Cottoning on

? Why do we prefer to use cotton for our clothing – a water hungry crop – and not synthetic fibres? Is it for particular characteristics that synthetics still do not have?

Jean Esnault
Angers, France

We love cotton clothing because it is comfortable, soft and easy to clean. Cotton is comfortable as it has a high amount of moisture regain, generally about 7 to 7.5 per cent. Moisture regain is the amount of moisture that the fibre will absorb, expressed as a percentage starting from a bone dry condition, when placed in an atmosphere of standard temperature and humidity. This means cotton can absorb our body moisture and give us the sensation of being cool.

Polyester, a synthetic that many fabrics are made from today, only has a moisture regain of 0.4 per cent, and will feel hot sticky and clammy, especially when the humidity is rather high. Cotton feels soft because it has a ribbon cross-sectional shape, and as a result will bend easily when pressure is applied. Most synthetic fibres have a circular cross section and tend to be stiff because of their low elasticity. In the last twenty years, methods have been developed that enable the manufacture of fibres that are very fine in diameter, and as a result many synthetics now also feel rather soft.

Cotton fabrics are also easier to clean. When our clothing becomes soiled, most people wash them with soap and water. Cotton is alkali resistant and therefore soap doesn't damage the fibre. In addition, cotton is stronger when it is wet, and so cotton clothing will resist being damaged by the agitation of the washing machine. Most important is the fact that cotton cleans well when washed. Polyester and olefin fibres tend to absorb our body oils and other oil-borne stains which are difficult to wash out. Cotton doesn't readily become stained with our body oils, and therefore will tend to remain cleaner.

J. Robert Wagner
Plymouth Meeting, Pennsylvania, US

Synthetic fibres get such a bad press that I thought I should mention some of their benefits. Cotton has poor abrasion resistance compared with synthetics such as nylon and polyester, and also shrinks more easily when washed. Despite advances towards solving these problems, major stores are still selling cotton items that shrink.

Cotton's advantage in absorbing water is also a disadvantage, in that more energy is required to dry it. While this may not be a problem in the summer, when garments can be dried outside, during the winter the extra heat required is significant. Making synthetic yarns is clearly an energy-intensive business. However, this energy may be less than the extra energy required to dry cotton, taken over the lifetime of the garment. Clearly this additional and hidden energy use should be considered when clothes are designed. Unfortunately, it is frequently overlooked by manufacturers, who go for the easiest option, with long-term implications for the climate.

My view is that a mixture of old and new fibres in some products would lead to better-quality clothing with reduced energy use.

Brian Bennett
Lathom, Lancashire, UK

Explosive power

? Last night my wife and I were awoken by two loud cracks a second or so apart. I discovered that the 9-volt battery cell I'd removed from a fire alarm appeared to have exploded. What might have caused this in a depleted cell, and why two cracks?

David Cross
Malvern, Worcestershire, UK

This is a 'battery' in the real sense, being constructed of six 1.5-volt cells, connected in series. Some batteries have stacked flat cells, but others have a three-by-two array of cylinder cells. Because the battery's terminals are at the same end, a circuit can easily be completed if they come into contact with a metal object – if stored in a metal box say. The battery has an internal resistance and heats up when a current flows through it. This causes the water in its electrolyte to vaporise and increase the internal pressure then . . . bang!

The first crack would have been the electron-collecting pin being fired from a cell like a bullet from a gun. The second would have been a failure in the casing of the adjacent cell, causing it to be ejected from the battery. Even a partially depleted battery can supply enough current to heat the metal contacts and internal parts and also any metal that the battery terminals may touch – and perhaps start a fire. There have been serious injuries to sensitive parts when 9-volt batteries have come into contact with coins in trouser pockets, completing a short circuit.

David Muir
Science Department, Portobello High School, Edinburgh, UK

Drying dilemma

? Which method of drying my hands uses less energy: an electric blow-dryer or a paper towel made of dried wood pulp?

Ian Cutter
Victoria, Australia

This depends on how the energy cost is measured. The effort of grabbing a tissue and wiping is less than it takes to generate a several-hundred-watt blast of hot germy air in which to rub your hands interminably. However, the energy expenditure of producing and recycling the paper is more difficult to assess. Wood had to be grown, it had to be harvested, transported, shredded, pulped, matted, dried, configured into saleable and usable formats and packaged. After use, a recycling plant had to re-pulp the material and repeat much of the cycle, which only in ideal circumstances is cheaper than producing new paper.

Then again how much energy did it take to manufacture the blower and ultimately to recycle it? How large a pile of consumed paper would the device have to eliminate to justify its role as an implement of conservation? Perhaps it would be best to carry your own hand towel or simply to dry your hands on your trouser legs. Towelling pads sewn onto clothing could become a conservation-conscious fashion statement to rival the breast-pocket handkerchief.

Jon Richfield
Somerset West, South Africa

Circuit judge

? During a blackout, does electricity reverse its flow from the battery of my phone to the charger or power outlet?

Roque Mansiliohan
San Luis, Philippines

Mains current is alternating, which means it reverses direction many times a second. This kind of current is easier to generate, and is necessary for most motors and transformers to work. At the heart of all chargers is a component called a rectifier. This blocks the flow of electricity in one direction. As well as blocking the backwards half of the mains cycle, it also prevents your battery discharging back into the mains. Modern phones contain battery-management hardware and software which prevents the battery from getting overcharged or charging too quickly. This software would shut down the charger if it tried to discharge the battery, or if the rectifier failed.

Philip Belben
Coalville, Leicestershire, UK

Surge protector

? You often hear that extra power stations have to be put online in the UK during ad breaks for popular TV shows because of everyone making a cup of tea. Is this really true, and if so does national power usage in the UK vary more than in other similar countries?

Simon Scarle
Newport, Gwent, UK

As demand for power rises, power companies do indeed bring extra generators online to cope. These companies would love to be able to hold electricity in batteries for such events as advert breaks, morning coffee and evening meal times, but there is no efficient way of storing the alternating current power stations generate, given that it needs to be converted to and from direct current for storage.

Demand for power is monitored and controlled on a minute-by-minute basis from a central control station. Sophisticated prediction software, using many years' worth of data (and the TV listings), helps to keep the system running as efficiently as possible. But even with this, unusual shifts in weather can affect how stations are controlled. For instance, a cold spell will affect demand because people turn on extra heating to keep warm. If there is less wind, more power stations will need to start up because wind generators are not able to work.

All countries similar to the UK have the same problems and in 1986 a 2,000-megawatt high-voltage link between the UK and France was installed. This enables the two countries to share electricity during demand peaks on either side of the channel. It saves the need to bring too many power stations online because demand peaks in France rarely coincide with those in the UK – except, that is, during half and full time of Six Nations rugby union matches!

The UK now also has similar connections with the Netherlands and the Republic of Ireland.

Paul Keddie
Llangollen, Denbighshire, UK

It's behind you

? Many years ago I heard a radio broadcast featuring a beeping sound that always seemed to come from behind me. The announcer said that the sound would have this quality and it did, even when I turned around. It was a plain beep, and the radio only had one loudspeaker. It was a complete mystery to me then, and it still is, so can anyone explain the effect?

Martin Harvey
London, UK

Many people assume that the outer ear is shaped as it is purely to collect sound and channel it into the ear canal. The shape certainly helps with this task, but the ear is also designed to subtly change the quality of the sound so that we can distinguish where it is coming from.

When a sound is coming from behind you, it enters the ear canal from a particular angle, causing it to pass over a specific part of the outer ear. We are not conscious of the way this changes the sound, but our auditory cortex can detect it. When combined with information about the amount of sound received by each ear, this enables us to tell where the sound is coming from.

In order to create a sound that constantly seems to be coming from behind you, a technique known as dummy head recording is used. A microphone is placed inside the ear canal of a model of a mannequin head with a realistically shaped outer ear. This alters the sound before it reaches the micro-phone, meaning that the position of the sound relative to the head is maintained in the recording. When it is played back, you perceive the sound to be behind you, regardless of where your radio is positioned.

Toby Bateson
Truro, Cornwall, UK

Ocean dwellers

? Has any thought been given to the idea of building cities on the sea? Surely the technologies exist to make this a realistic alternative to destroying more agricultural land, and it could even allow us to benefit from sea level rise.

Peter Hoare
Ashwicken, Norfolk, UK

The idea of building cities on the sea has been around since at least the early 1960s. Buckminster Fuller designed a tetrahedron-shaped floating city intended for Tokyo Bay. Only now is it close to becoming a practical reality.

PayPal co-founder Peter Thiel is one of the backers of the Seasteading Institute, which plans to get a modular floating city costing $170 million on the water by 2020. There has also been publicity about French architect Jacques Rougerie's 900-metre-long manta-ray-shaped floating city that would house 7,000 people. Another concept is Vincent Callebaut's lily-pad-shaped city that could house 50,000.

The main issue holding back all of these ideas is the amount of money required from investors to build them. Until this funding is secured, we will have to be satisfied with huge cruise liners that have almost become floating cities themselves.

Peter Hassall
Wellington, New Zealand

Unsuitable suit

? I read that Neil Armstrong's spacesuit from his moon landing is deteriorating. It was, apparently, made out of materials created for only short-term use. Why was this, and what are these materials? How can the suit be preserved? The spacesuit of another Apollo astronaut, David Scott, is on display in the Smithsonian National Air and Space Museum in Washington, D.C. This suit looks pretty much intact, so was it made of different materials?

Alan Brewer
Washington, D.C., US

One simply can't go into space without wearing a spacesuit. Even Laika (the first dog to orbit the earth) wore a bespoke suit. Her outfit never made it to a museum, although copies exist.

All used American spacesuits, including prototypes, spares and those stained with moon dust, are in the care of the Smithsonian Institution. Most are kept in storage, some are on loan for display. All are degrading; those on display are degrading faster, some of them at alarming rates.

My expertise is in maintaining environmental conditions suitable for museum storage and display. As a result, I have had the pleasure of helping to protect some of the many artefacts and spacesuits in the Smithsonian's collection. This means I can answer at least some of the question asked above.

Spacesuits are made from the most effective materials for the job. Strength, weight, impermeability, flexibility and other variables are judged and balanced when a spacesuit is designed and produced. If it protects its wearer from the dangers of space, enables movement and is comfortable, it has done its job. Long-term viability is not likely to be a matter of concern: suits are made to withstand the severe rigours of space, not to survive the dangers of long-term museum display.

The body of a spacesuit is usually comprised of multiple layers of plastic-based cloth and barriers, often bonded or fused together. The rest of the suit is a combination of metals, plastics, wires, insulators, tubes, anchors, glues, gasketing, lubricants, paints and inks. Not only will many of these materials degrade spontaneously, but they may interact in unexpected ways over time. Heat and humidity changes will hasten their degradation, as will airborne pollutants and light. The effects of poor storage conditions may not become evident for years.

While not on display, the Smithsonian's spacesuits are kept in the dark, under constant cool and dry conditions. When the suits are on display, efforts are made to protect them from light, dust and less-than-suitable environments. To this end, the new Milestones of Flight gallery in Washington, D.C.'s National Air and Space Museum will incorporate positive pressure microclimate environmental control systems for any sensitive items, including space-suits. Airtight showcases will be supplied with a constant flow of scrubbed and humidified air by a microclimate generator. While this will not stop degradation, it will slow it.

Jerry Shiner
Keepsafe Microclimate Systems, Toronto, Canada

Right on Q

? I use predictive text on my cellphone. When I type the letter 'q' I get suggestions all of which start with 'qu'. When I then press 'u' I get different suggestions. Why? I realise it's quite trivial but I'd still love to know the answer.

Andrew Barnett
Cheltenham, Victoria, Australia

When you type the letter 'q', some words are suggested by the phone. If one of these words was the word you wanted to type then the phone would expect you to select that word from the list, rather than keep typing. So when you follow by pressing the letter 'u', the phone assumes that the word you are typing hasn't yet appeared in the list, and so it suggests some different words.

Catriona Quin
Dunblane, Perthshire, UK

10 What the Heck is This?

Eggs-tra shell

? I was shopping yesterday and found little clusters on the surface of an egg. They look like extra shell or a calcium build-up, but if they are what causes this? If they are deposits of extra shell, why are they formed into these strange clusters rather than spread evenly over the surface so that the whole shell is thickened?

Charlene Timberlake
Portsmouth, Hampshire, UK

Such deposits are indeed made of calcium which forms the hard outer layer of the shell as the egg passes through the shell gland. The deposition of extra calcium can occur for a number of reasons, the most common of which is the egg spending longer than usual in the shell gland. This normally occurs owing to stress, for reasons such as changes in lighting, habitat, or diet, or the presence of predators or foreign species.

Lack of water also induces stress and leads to the shell being deposited less evenly, because the deposits are not properly diluted. A further cause of irregular shell formation can be infection by diseases such as bronchitis or encephalomyelitis. Diseases like these can of course alter many aspects of the chicken's metabolism and biological activity, including egg formation.

Lewis O'Shaughnessy
Salisbury, Wiltshire, UK

Eggcellent question . . . When the egg is forming in the hen's reproductive tract it spends most of its time in the uterus. The uterus contains the shell gland, which lays down calcium carbonate around the outer membranes of the developing egg. While in the uterus, the egg rotates and the shell is not always laid down or deposited in a uniform manner. So the bubbles you see on the exterior of the egg are extra calcium carbonate from the shell gland. As hens get older, their shell glands may not work as well, so sometimes you see small abnormalities like those in the question. This can be due to age, the shedding of cells, or viral infections. If you put an egg up against a bright light you can see the non-uniformity of apparently normal eggshells. This mottling is the result of air and shell being laid down in a non-uniform manner.

Maurice Pitesky and Rodrigo Gallardo
UC Davis School of Veterinary Medicine, California, US

Foam home

?︎ In winter I visited Tywyn in west Wales. Both the beach and the promenade were covered in a mass of foam blowing in from the sea. The sea was rough with waves driven by a strong wind, and the foam covered the surface. The next day conditions were similar but there was no foam. What caused it and what changed overnight so that it disappeared?

Jim Logan
Gatehouse of Fleet, Dumfries and Galloway, UK

Sea foam or spume is created as seawater is churned when waves break on the shore. It is more likely to happen when the water contains dissolved organic matter, such as when algal blooms decompose. Compounds like these act as foaming agents, or surfactants, which trap air, leading to the formation of long-lasting bubbles when the water is whisked by wave action.

Wind blowing across the oceans can create foam-tipped waves, or whitecaps, in a similar fashion. Assuming average sea-level wind speeds, whitecaps should cover between 2 and 3 per cent of the oceans at any one time – about the same area as the US.

According to climate scientists, a warmer world should also be windier, increasing the percentage cover of whitecaps. As white surfaces increase the amount of sunlight reflected back into space, this could provide a brake on warming. Indeed, an increase in wind speed from 7 to 9 metres per second is thought to triple the percentage of sea surface covered by foam.

Mike Follows
Sutton Coldfield, West Midlands, UK

Beach balls

 What causes soft little balls of beach sand to form? I saw huge numbers of them stretched along about seven hundred metres of a beach about a hundred kilometres north of Auckland.

David Goldkorn
Auckland, New Zealand

They could be the casts of the lugworm or sandworm *Arenicola marina*. The small sandy heaps typically looking like miniature coils of cable are usually between 1 and 3 centimetres tall. The casts often cover large areas of sandy beach at low tide. The worms make excellent bait for all sorts of fish and crabs. Fishers looking for them find their casts to be useful indicators, because the worms themselves lurk between 10 and 30 centimetres beneath.

As a child I lived on the Isle of Wight off the south coast of the UK and I used to accompany my father on digging trips to the area on the east side of the pier at Ryde, an excellent spot for finding lugworms. In certain tidal conditions, when constant currents are moving slowly over shallow areas, some of the mounds of coiled sand cast off by the burrowing lugworms can become dislodged from the seabed. Then, rather than simply disintegrating, they form into small balls or beads of sand that are left behind when the tide recedes. This happens particularly where the sand is slightly oily, such as when the beach is contaminated with pollutant-covered debris washed up on or near the shoreline.

Anthony Wilkins
Sowerby Bridge, West Yorkshire, UK

The prime suspect would be what are commonly called sand bubbler crabs, a name that covers animals which can belong to the genera *Scopimera* or *Dotilla*. After these crabs have sifted through a batch of sand for microscopic food, they discard it as balls about half a centimetre across. That said, I would expect the balls to be more densely packed around the crabs' burrow, but the photo suggests a fairly even distribution. Moreover, although the crabs are widespread in the Indo-Pacific region, New Zealand may be just beyond their range.

Mike Follows
Sutton Coldfield, West Midlands, UK

Shell suits

? I found an egg among my groceries that had a strongly ridged shell with the pattern of ridges resembling veins or arteries, while the surface between the ridges was densely pitted with a few isolated tiny pimple-like protrusions. The deformities are more prominent on the narrower end of the egg. What is the explanation, and is such an egg safe or palatable to eat? I did not feel inclined to put this to the test.

Brian Bonney
Southampton, UK

The most likely cause of the deformity is an interruption to the part of egg formation called the plumping process. An egg begins to form when the ovum (the yolk) is released from the hen's ovary. It is then picked up by the infundibulum and proceeds on down through the reproductive tract picking up albumen (the white) as it goes. After the shell membrane is added, a process called plumping occurs in which fluids pass through the membrane, causing the white to swell and fill the space within the membrane. Once full, the shell hardens. If something goes wrong with the timing of this process, the shell may harden before it is completed. In that case you get an egg which may be misshapen in the way the photograph shows. It is surprising, however, that such an oddly shaped egg reached the consumer market, because they are usually rejected during the grading process.

Kim Critchley
Clarence Park, South Australia

Trail of mystery

? In my grandmother's house a snail-trail-like substance is visible on the mat in the conservatory each morning. There are no apparent signs of a snail living in the room and the door is locked at night. Is there another insect or animal that causes this effect?

Shane Duffy (aged twelve)
Dublin, Ireland

The snail-trail-like substance is indeed caused by a snail-like creature but one without a shell – the slug. Having recently been on a course run by the National Museum of Wales, I have begun to appreciate the subtleties of a slug's life. It is more than likely that the slug is living in your grandmother's house under a skirting board or cupboard, having probably entered attached to someone's shoe. In light, it will immediately seek darkness and only appear at night.

The slug or slugs will be delighted if there is a plate of dog or cat food close by. The slime trail that the slug leaves is one of its greatest defences, because most animals, including humans, don't like touching them, and so leave the creatures alone. However, it is also a weakness because it reveals that the slugs have been present. There are many different species of slugs found in the UK, and the 2014 book *Slugs of Britain and Ireland* allows you to identify any slugs you may find, and discover whether they are a garden pest (as many keen gardeners would have you believe), or helpful for example with composting. I am saving my copy for when my grandchildren next visit so that we can all go on a slug hunt and discover how many species I have in my garden – or my house.

Gillian Coates
Trefor, Anglesey, UK

Mississippi melter

? I found an odd thing on a sand beach along the Mississippi river in Minneapolis in early September. It was solid and it had the consistency of a jellyfish, but appears to have formed along a strand of reed, seen running through the middle. Can anyone explain what it is?

Andrew Koebrick
Minneapolis, Minnesota, US

This sounds like *Pectinatella magnifica*, a member of the Bryozoa – a group commonly known as 'moss animals'. The gelatinous mass is in fact a colony of dozens of animals, or zoids. A popular name is jelly ball.

Most colonies are elongated and they adopt the best shape for clinging to the object to which they adhere. Free-floating colonies tend to be spherical. They are found in stagnant or slow-moving water, so this one in the Mississippi must have floated in from a tributary or a riverside pond. I once encountered a round one as big as a soccer ball. They sometimes sprawl out over a metre or so, resembling a giant amoeba.

Donald A. Windsor
Norwich, New York, US

Odd jobs

[?] I have recently seen some strange objects in three places near my home – twice on verges of country roads and once near a stable's muck heap. They are small spheres between 2 and 4 centimetres in diameter and very hard. They are nearly black inside with terracotta-coloured skin. Each site had several hundred of the objects in a rough heap about a metre across. Have they been manufactured for a purpose? If so why are they being dumped? Or are they natural? If so what created them?

John Secker
Daresbury, Cheshire, UK

These are expanded clay pellets or aggregate. They are produced by firing natural clay at high temperatures in a rotating kiln, to produce spherical pellets with a terracotta-like surface. A black porous core is formed by burning organic materials within.

Aggregate is manufactured for many purposes including for use in building materials such as blocks slabs and concrete. It is useful for this because the clay pellets are lightweight and are excellent insulators of sound and heat, and they are resistant to rot and chemical degradation. It is also often used in building situations where loads need to be minimised, such as in the substrate above a tunnel where the pellets also provide stability and efficient drainage.

Additionally the pellets can be used for water treatment due to their high surface area which aids the absorption of pollutants. The ones shown have most likely been used as a growing medium and discarded. The clay allows for the absorption and retention of water as well as allowing the drainage of excess water from around the roots of whatever is planted in them. This ensures that air circulates throughout the medium and the porous pellets maximising oxygen uptake. They are often used with a constant or regular water supply, such as a drip feed or an ebb and flow system. The pellets can maintain a neutral pH as well as providing a large cation-exchange capacity for the retention and provision of plant nutrients added to the water.

Because they have been left in relatively small heaps by the road it's not unreasonable to suggest that they have been used as a growing medium for the cultivation of cannabis, dumped within a rural environment to avoid suspicion.

Patrick Melia
Brighton, UK

11 A Last Word From Us

The *New Scientist* inbox offers a never-ending supply of questions from our readers, and their curiosity about the world is matched only by our own. Collected below are twenty-three questions we at *New Scientist* have found ourselves asking at one point or another. Some were prompted by current events, while others came out of nowhere. And some are ones we've been asked so many times, we had to find out the definitive answer once and for all. So here they are. We hope you enjoy them. And if you think we've missed something, send us your own answer – you may just end up having the last word on the subject.

Wandering minds

? Can you improve your attention span? And how long should it be, anyway?

Our wandering minds tend to make most of us feel guilty. 'We have anecdotal evidence that most people think they mind-wander more than average,' says Jonathan Schooler of the University of California, Santa Barbara. Schooler has attempted to assess 'normal' levels of mind-wandering in the lab by, for example, getting people to read extracts of Leo Tolstoy's *War and Peace* and interrupting them to ask their thoughts at random intervals. Such studies reveal we spend anywhere between 15 and 50 per cent of the time with our head in the clouds.

Such a lack of focus might seem terribly inefficient, but probably isn't. 'It's unproductive in the context of whatever you are doing currently,' says Schooler. 'However, it is potentially productive in the context of whatever it is you're thinking about. You might be reading a book and thinking about planning a party, and it's totally compromising your ability to read the book, but you're making progress on the party.' There is good evidence that a wandering mind is an evolved trait that helps us to think about and plan for the future – something that also fosters a uniquely human creativity.

Even so, might there be too much of a good thing? One common worry is that our concentration is dwindling as technology provides more distractions. A recent report published by Microsoft claimed that the average Canadian attention span was down from 12 seconds in 2000 to 8 seconds in 2013 – one second less than a goldfish.

J. Bruce Morton at Western University in London, Canada, is sceptical. For one thing, there's no standard measure of an attention span, he says. 'What people want to hear is, how long can you concentrate for? And what should a child's attention span be? But those are not concepts that are normally used amongst scientists studying the topic.'

Measuring perhaps the closest thing – the ability to stay focused on one task, known as selective attention – involves looking at attention shifts on the millisecond scale, such as asking people to state the colour of shapes as they pop up on a screen while ignoring distractions that pop up at the same time. Such experiments show a lot of variation in selective attention. It's low in kids, perhaps because the developing brain has yet to master control over areas that process incoming sensory information. It then improves until the age of twenty, when it plateaus until middle age, before diminishing once again.

But there's no evidence technology is making us worse at concentrating, says Morton. Instead, it has just become so well designed and intuitive that it takes advantage of our innate ability to think of several things at the same time.

If you are still worried, there are things you can do. Stay off the booze – alcohol reduces our awareness of mind-wandering, says Schooler. 'When you consume alcohol, you mind-wander more and notice it less.' Technologies that promote thought control could help, as could meditation. 'People who practice mindfulness mind-wander notably less,' says Schooler.

Catherine de Lange

A whale of a job

? Why do whales strand themselves, and what happens next?

Strandings are more common than you might think: the UK alone sees almost six hundred cetaceans washed up on its beaches each year. These are mostly dolphins and porpoises, but around fifty are live whales. Sometimes strandings have a clear link to human activity: whales wash up after being hit by ships or tangled in fishing lines or nets. Pollution and marine noise have also been implicated in mass strandings. But often the cause is unknown.

Sperm whales usually live in the deep ocean, where their main prey, squid, is plentiful. If they get into shallow water, like the North Sea, they are in trouble. Food is harder to come by, so they get hungry and dehydrated, because all of their water comes from food. Being in the shallows also impairs the sonar that whales use for navigation, possibly making them more confused and more likely to beach.

If that happens, the prospects are not good. Whales' enormous mass is normally supported by water. When they are beached, the weight of their bodies damages their internal organs and muscles. The protein myoglobin is released from their muscles into their blood and reaches their kidneys, where it is highly toxic. Dehydration and kidney damage are the most common causes of death.

Once whales have been stranded for an hour, the damage to their kidneys is irreversible. Getting them back into the water doesn't guarantee their safety, as their blood can circulate more freely, and this carries even more myoglobin to the kidneys. In 2002, volunteers managed to refloat forty-five pilot whales that washed up on Cape Cod in Massachusetts. But the next day, they beached again, and those still alive had to be euthanised.

One group of experts, the Marine Animal Rescue Coalition, says trying to rescue stranded whales is usually futile, and it's kinder to euthanise them. But this is far from simple for such a large animal. Here in the UK, we just don't have a big enough supply of drugs to do that routinely, and we have to let nature take its course.

After they die, the whales start to decompose, producing a stench that attracts scavengers, and causing gas to build up inside their insulating blubber. Sometimes the pressure gets so high that they explode. According to British law, whales that wash up on the coast are the property of the crown; the head belonging to the king and the tail to the queen. In practice, scientists working for the UK Cetacean Strandings Investigation Programme get first dibs, and can decide to carry out a post-mortem. Once they're finished, it's up to the local council to dispose of the carcass. Usually, they are cut up and taken to landfill, rendered or incinerated. East Lindsey District Council in Lincolnshire estimated that it cost about £10,000 to dispose of a beached sperm whale in 2012. In some countries, beached whales are towed out to sea or blown up.

Could a better use be found for them? If you're thinking of eating the leftovers, it's not recommended. Whale meat and blubber have been found to contain high levels of mercury, botulism toxin and even pesticides. It's not a meal fit for a queen – or anyone else.

Sam Wong

Wheel mystery

? How exactly does a bike stay upright?

In 2011, an international team of bi-pedal enthusiasts dropped the bombshell that, despite 150 years of analysis, no one knows how a bicycle stays upright. Across the world, riders dismounted and stared at their bikes in disbelief. What they had been doing for years was a feat inexplicable by science.

Well, sort of. 'What we don't know are the simple, necessary or sufficient conditions for a bicycle to be self-stable,' says Andy Ruina, an engineer at Cornell University in Ithaca, New York. We have relied on trial-and-error engineering to construct stable bikes that aren't prone to toppling while in motion. Explaining how they work mathematically requires around twenty-five variables, such as the angle of the front forks relative to the road, weight distribution and wheel size.

Before 2011, researchers had reduced this profusion to two things. One was the size of the 'trail', the distance between where the front wheel touches the road and where a straight line through the forks would meet the ground. The other was the gyroscopic restoring force that acts on a spinning wheel to keep it upright.

Ruina and his colleagues, including Arend Schwab of the Delft University of Technology in the Netherlands and Jim Papadopoulos of the University of Wisconsin-Stout at Menomonie, not only revisited this mathematics, but also skewed the trail and gyroscopic forces in prototype bikes to make them technically un-rideable. To everyone's surprise, the bikes were still stable.

The researchers haven't been resting on their saddles since. Last year Ruina unveiled a 'bricycle', a cross between a bicycle and a tricycle with spring-loaded stabilising wheels that can be adjusted to vary the rider's perception of contact with the ground. By studying the influence this has on how the rider steers and remains stable, he hopes to gain new insights that might lead to more easily controllable bicycles.

It's still an uphill struggle. 'I think the real understanding of bikes requires a mix of what we did, plus some kind of brain science,' says Papadopoulos. Human riders act in extremely complex yet intuitive ways to keep a bike balanced and on track. At very low speeds, for example, we recognise that the handlebars become useless for steering, and instead direct the bike by wobbling our knees.

Why? 'We don't know,' says Schwab. Yet another bike-based mystery that could be around long after we've worked out the origins of the universe.

Michael Brooks

Astronomical error

?　Why isn't Pluto a planet?

On 14 July 2015, NASA's New Horizons probe beamed back a now-iconic picture of Pluto, revealing its craggy mountains, deep valleys and heart-shaped ice plain in greater detail than ever seen before. 'How about a round of applause for that beautiful planet?' beamed mission leader Alan Stern as he presented the photo to an adoring audience.

Thing is, Pluto isn't officially a planet. If there were any doubters in the room, they were too polite to contradict Stern in his moment of glory, but since 2006 this distant world has been classified by the International Astronomical Union (IAU) as a dwarf planet, meaning it no longer ranks among the 'proper' planets of the solar system.

That's puzzling. As New Horizons clearly demonstrates, Pluto looks like a planet – it passes what Stern likes to call the *Star Trek* test; you can imagine the crew of the *Enterprise* looking out from the bridge and deciding to land there. So why did Pluto get kicked out of the club? To understand that, we need to roll back to the very beginnings of astronomy.

The word planet comes from the ancient Greek term *asteres planetai,* or 'wandering stars'. Even before the invention of the telescope, early civilisations noted there were seven astronomical objects that didn't remain fixed in the cosmos: the sun, the moon, Mercury, Venus, Mars, Jupiter and Saturn. With the Copernican revolution, astronomers realised we revolved around the sun, not vice versa, and swapped out the sun and the moon for Earth, giving us six wanderers.

Uranus was the first new planet discovered with the aid

of the telescope in 1781, followed by Ceres, Pallas, Juno and Vesta. If you're not familiar with these, that's because we now call them asteroids – they are all small, mostly lumpy space rocks that orbit between Mars and Jupiter – but for the first half of the nineteenth century, they were ranked alongside the larger planets, for a total of eleven. The discovery of Neptune in 1846, along with dozens more small bodies, forced a rethink, as it became clear these asteroids were altogether different.

Enter Pluto. As soon as it was discovered in 1930, astronomers could see it was an oddball world, orbiting at an angle to the other planets and occasionally crossing in front of Neptune's orbit. But Pluto was welcomed to planethood with open arms. The solar system was complete.

Or so we thought. In the 1990s Pluto's status wobbled as astronomers began discovering trans-Neptunian objects (TNOs), a group of smaller bodies in a similarly distant locale, known as the Kuiper belt. Then in 2005 came the crushing blow: the discovery of Eris, a TNO that seemed to be a hair larger than Pluto.

Astronomers faced a crisis. Should Eris be anointed as the solar system's tenth planet? What if other would-be planets were lurking out there, ready to tear up textbooks at a moment's notice? The IAU was forced to react, and at a meeting in August 2006 it demoted Pluto to dwarf planet status – just months after the New Horizons mission had been launched to explore it.

Planets and dwarf planets are both round objects that orbit the sun, but dwarfs lose out on full rank because they have not hoovered up or displaced other smaller bodies from the neighbourhood around their orbit. Pluto falls foul of this clause due to other nearby TNOs, but Ceres gets an upgrade from mere asteroid to dwarf planet, since it is the only round object in the asteroid belt. The dwarf planet definition also explicitly excludes moons, otherwise our own moon, along with many others in the solar system, would qualify.

And yet, when you look at pictures of Pluto, you see a planet. The same is true of Ceres, which got its own close-up in 2015 thanks to NASA's Dawn probe, the asteroid's first visitor. If the *Star Trek* test – roughly that the object is round and has interesting features – is good enough for Captain Kirk, why not us as well?

For one thing, that definition would put the sun back in the planet list, as the ancients once had it, so we had better make orbiting our sun a condition of planethood, as the IAU did. That would mean the solar system has planets in double figures, given the dozens of small, round TNOs out there – some of them even have their own moons.

But hang on – moons clearly pass the *Star Trek* test, especially as our moon is the only other world we've set foot on. And Triton, the largest moon of Neptune, around 300 kilometres wider than Pluto, is thought be a former TNO that got captured by the gas giant. Should planets really become moons simply because they have a new address? It's a thorny and undecided issue.

And these days we've got more than the solar system to worry about. Telescopes like NASA's Kepler spacecraft have now racked up over 3,000 exoplanets orbiting distant stars, but apart from a few very large ones, most of these have not been photographed directly. It's very likely all known exoplanets are round, as they seem to be massive enough that we'd expect them to self-gravitate into spheres, but future telescopes could find smaller, more ambiguous worlds. Could we only call them planets after launching interstellar probes?

The IAU's planet definition actually has nothing to say about exoplanets, but a separate definition decrees they should be below thirteen Jupiter masses, to distinguish them from brown dwarfs, a kind of failed star that straddles the planetary line. This definition is controversial, and also excludes free-floating 'rogue' planets that have no star. Again, the tensions between form and location make it tricky to pin down a planet definition that everyone can agree one.

Ironically, New Horizons' detailed measurements of Pluto's radius confirmed that the dwarf planet is actually just larger than Eris, so the argument that erupted after the 'tenth planet' discovery might have been avoided if only we'd started exploring sooner. Stern now likes to talk of the Kuiper belt worlds as the 'third zone' of the solar system, after the rocky planets and gas giants. In a lecture on the eve of New Horizons' arrival, he argued we should embrace their planethood because they are actually in the majority. 'It's really a third class of planet in the solar system,' he said. 'Just like insects outnumber humans, the small ones are winning.'

Jacob Aron

Lights out

Is losing consciousness like a light snapping off, or moving a dimmer switch?

Consciousness feels like an on–off phenomenon: either you're experiencing the world or you're not. But finding the switch that allows our brains to move between these states is tricky. 'Consciousness is not something we see, it's something through which we see, which makes it challenging to study,' says George Mashour, director of the Center for Consciousness Science at the University of Michigan in Ann Arbor.

One common definition of consciousness is 'the thing that abandons us when we fall into a dreamless sleep, and returns when we wake up'. But say I anaesthetise you: you may hear my voice, but not respond to it; you may be dreaming and not hear my voice; or you may hear or experience nothing at all. What patterns of brain activity correlate with these levels of conscious experience?

We do know there are certain brain regions that, if damaged or stimulated, cause loss of consciousness. The claustrum – a thin, sheet-like structure buried deep inside the brain – is one of them. But many leading theories that aim to describe consciousness veer away from a single anatomical site being its seat.

Global workspace theory hinges on the idea that information coming in from the outside world competes for attention. We only become conscious of something – a ringing telephone, say – if it out-competes all else to be broadcast across the brain.

Then there's information integration theory, which suggests that consciousness is the result of data being combined to be more than the sum of its parts. 'Unconsciousness doesn't necessarily need to be mediated by brain regions being shut down or extinguished, but rather by a communication breakdown,' says Mashour.

A recent study that scanned people's brain activity as they were slowly anaesthetised seems to back up that picture. It might also explain how a drug like ketamine knocks people out: this potent tranquilliser ramps up activity in many of the brain areas that promote wakefulness, but depresses communication between different regions.

Linda Geddes

Moon shot

⟨?⟩ It's often viewed as a crowning achievement of human exploration – yet it last happened more than forty years ago. Why haven't we been back to the moon?

Do you want the silly reason? The political reason? Or the one that takes a bit longer? The first is the stuff of conspiracy theorists: that we never went in the first place and don't have the ability to do so. This is baloney, naturally. The second reason, at least, is true. The race to the moon between the Soviet Union and the ultimate victor, the United States, was driven by ideology forged in the Cold War. Two sides, peering at each other over their nuclear warheads, sought to show that their politics would prevail. Putting a person on the moon would somehow prove that unequivocally. Once one of them had, there was no imperative to go again, which is why interest in the moon landings petered out when the programme was cancelled after Apollo 17.

But nations still crave grandeur and kudos to prove their worth, which another moon landing – in the watch-every-event-on-your-smartphone era – would certainly achieve. There's no shortage of candidates – one of the fading European imperial nations, or China, India, Brazil, even North Korea. So why aren't they going?

This is the reason that takes a bit longer. Could we still go to the moon if we were willing? To a certain extent, we have lost the ability to do so. Imagine the massive investment of time, money and technology involved in taking Neil Armstrong and Buzz Aldrin to the Sea of Tranquillity. The huge Saturn V rocket that blasted them there, the launch tower, the phenomenally complex design of the lunar landing module, the suits, the heat shield, the parachutes . . . once you stop going to the moon, you really don't need these things any more. So you don't keep the factories that built these vehicles and their accoutrements, you close them down. Job done.

So if you wanted to go again, you would have to start from scratch. New designers (the old ones are nearly all dead), new designs, new developments in aeronautics and rocketry, new materials, new factories, new launch pads, new absolutely everything including, most importantly, computers.

We could, of course, retrace our steps, just as we could rebuild a Colosseum exactly as the Romans built theirs two thousand years ago. But why would we? Our stadiums are constructed completely differently today. And why aren't we still using Concorde or its successor to cross the Atlantic in half the time? Progress isn't necessarily linear. For a new moon shot we would use computers, cranes, trucks, materials and lasers that didn't exist in the 1960s. Although it would be more efficient – we'd be foolish to use the old technology – we'd be going back to the drawing board.

And the goals would be different this time, too. It's not about proving one's political system is superior, the public wouldn't hand over taxes for a vanity project any more. Apollo might have been justifiable in the heat of the Cold War, but today a moon programme would have to have serious scientific, social, commercial and environmental benefits before the public would buy into it. So it takes more than building a rocket – we would need to build a whole movement.

Finally, there is the spacecraft itself. It might have to be very, very big to carry this new payload of scientific benefit we're talking about. And that spaceship would have to be tested and tested before putting it on a course for the moon.

It can be done, with enough time, enough effort and

enough money. Even without political will, there are private firms already involved in spaceflight – although going to the moon would need to offer them some hugely beneficial return. Given the right commercial conditions, perhaps they would be interested. Until that day arrives, Eugene Cernan's title of last man to have walked on the moon will remain intact.

Mick O'Hare

Ultimate ancestor

? What was the first life on Earth?

In the beginning was Ida, the initial Darwinian ancestor – the first material on Earth to transform from inert to, well, ert. Ida begat Luca, the last universal common ancestor, a molecule that stored information as genetic code, and gave rise to all life on Earth.

Ida and Luca live on within us. Our cells all use the same genetic code embodied in DNA, suggesting Luca was itself made of DNA. Except it isn't that simple. All life uses proteins to make DNA and execute its code – but proteins themselves are made from DNA templates. Which came first?

Probably neither. RNA is a close relative of DNA found in all living cells that also carries genetic code and, crucially, can catalyse chemical reactions on its own. The RNA world hypothesis says Luca was born out of an RNA soup that eventually gave rise to DNA and the first cells.

But where did RNA come from? In the 1950s, American chemists Stanley Miller and Harold Urey famously zapped a mixture of gases and water with electricity and ended up with a handful of biotic molecules. Nowadays, though, more nuanced ideas are in vogue. Nick Lane of University College London, for example, thinks that warm vents on the ocean floor provided a soup of methane, minerals and water from which RNA could form. Michael Yarus of the University of Colorado in Boulder, meanwhile, favours a slushy pond whose continual freezing and thawing pushed chemicals together in just the right way.

Intriguingly, more recent experiments trying to coax RNA into existence have shown that when the chemistry is just right, many of life's building blocks seem to form almost spontaneously – increasing the likelihood it happened elsewhere, too.

Catherine Brahic

Glottal stop

? The steady advance of technology and globalisation mean that a few languages dominate and grow, while many others decline. How far will this this trend continue – will we all speak the same language one day?

With over a billion native speakers, Mandarin Chinese is the language spoken by the greatest number of people. English comes third, after Spanish. But unlike Mandarin and Spanish – both spoken in more than thirty countries – English is found in at least a hundred. In addition to the 335 million people who speak it as a first language, 550 million cite it as their second. It dominates international relations, business and science. All this suggests that English is on course to become the planet's lingua franca. It just probably won't be the English that native speakers are used to.

Millions of second-language English speakers around the

world have created dialects that incorporate elements of their native languages and cultures. These varieties are known as similects: Chinese-English, Brazilian-English, Nigerian-English. Taken together, they – not American or British English – will chart the language's future path.

Jennifer Jenkins at the University of Southampton, UK, describes how language experts previously believed that the world would follow one of two possible futures. Either everyone would end up speaking an Americanised version of English, or English would separate as Latin did before it, giving birth to new descendants. But neither of those is happening.

Instead, English similects are probably here to stay. Even in a future where China, India and Nigeria are global superpowers, English is likely to be the language of choice for international discourse, simply because it is already installed. Weirdly, this is not such good news for native English speakers. When all the people of the world have English, it's no longer a special thing, and native speakers lose their advantage.

They could even be at a disadvantage. Non-native speakers are all tuned to each other's linguistic quirks. 'If you put a Chilean, a Japanese and a Polish person in a discussion in English, they understand each other perfectly,' says Jenkins. 'Put one with two native English speakers and there might be problems.'

My own brand of English – a mash-up of transatlantic Anglo and deep rural Ireland – has helped me through many a scrape. My mother and father could never understand Tim Teighe, our dairy farmer neighbour in the early 1990s in Cork, so I acted as a translator. Whether 'just an accent' or a similect, my English certainly makes it hard to place me. Most people think I'm from Canada.

In time, English similects may begin to blend over national borders. New dialects are likely to form around trades or regions. These common goals will drive the evolution of the lingua franca, regardless of whether we call it English or not. That is not to say that all other languages will vanish. German will remain the language of choice within German borders. Even Estonian, spoken by just 1 million people, is safe. Likewise, the language directly descended from Shakespeare's English has staying power with Brits and Americans. But English, like football, will soon move outside their control, pulled into something new by the rest of the planet.

Hal Hodson

Sotto voce

> **?** Where does the voice in your head come from?
> And how do you know if it's normal?

For Socrates, it came as a warning when he was about to make a mistake. For Sigmund Freud, it was a loved one accompanying him when he travelled alone. Hearing voices has a long history.

And as those distinguished gents perhaps attest, it isn't always a sign of madness: our everyday thoughts often sound pretty voice-like. In 2011, Charles Fernyhough and Simon McCarthy-Jones of Durham University, UK, found that 60 per cent of us experience 'inner speech' with a back-and-forth conversational quality.

So where does inner speech end and hearing 'outside' voices begin? One answer is that an inner voice 'sort of feels like you', says Fernyhough, so you feel more control over it – but given how involuntary many thought processes seem to be, that is rather unsatisfying. 'The question is at the heart of the puzzle of hearing voices, and why we haven't got better at understanding it,' says Fernyhough.

On the back of their biggest study so far, he and his colleagues estimate that between 5 and 15 per cent of us hear outside voices, even if only fleetingly or occasionally. About 1 per cent of people with no diagnosis of mental illness hear more persistent, recurring voices. Around the same proportion of the population is diagnosed with schizophrenia, challenging the assumption that the two are related.

So far, there seems to be little difference between the brains of those who haven't been diagnosed with mental illness, but do hear voices, and those who don't hear voices. It's probably best to ask yourself one question before getting worked up about the voices in your head, says Fernyhough: are they bothering you?

Voices aren't the only expression of our inner thoughts – our minds tell us stories, too. This 'confabulation' is a symptom of some memory disorders, whereby people have false recollections. But the rest of us do it too. Experiments show, for instance, that when people are forced to make a random decision they later invent a narrative to explain it.

One theory is that this helps us make sense of a world that bombards us with information, and gives conscious rationale to decisions we make for unconscious reasons. Robert Trivers, an evolutionary biologist at Rutgers University in New Jersey, thinks our lies are more self-serving: by lying to ourselves, we lie better to others too.

This may explain the phenomenon known as positivity bias, whereby people overestimate their virtues. 'We put ourselves in the top half of positive distributions,' says Trivers. 'Eighty per cent of US high school students believe they are in the top half for leadership ability.' With these boosting voices, you probably shouldn't be too worried about what you hear – just don't believe everything they tell you, either.

Catherine de Lange

Galactic engineering

? Could we build the technology to change the galaxy?

Humans have begun to transform the Earth. We have built cities, transport networks and power stations, and sprinkled the skies with satellites. If we extrapolate this ability to engineer our own environment – for living space, travel, energy and communications – where does it lead us? Could we transform space too?

Predicting the far future is a fool's game. So let's take the usual dodge and say: unless something is forbidden by known physics, it will be done. Eventually. Before we start, let's invent two things: self-repairing AI supervisors that can direct projects lasting many millennia; and vehicles that can reach close to the speed of light, maybe riding on laser beams or driven by miniature black holes – which according to recent calculations by physicists at Kansas State University may be possible.

Thus tooled up, we could hop from solar system to solar system, sweeping across the galaxy in 10 million years or so, and then out into our local supercluster of galaxies. So the potential building site is quite spacious. Such a civilisation would consume lots of energy, and that is where our engineering may be most conspicuous. One option will be to plug in to local power sources, such as harnessing starlight with orbiting solar power stations. As demand for power grows, lots of these could be arranged to obscure a star completely, forming a closed 'Dyson sphere', named after physicist Freeman Dyson, who pointed out that technological civilisations tend to use ever-escalating quantities of energy.

If we built one we would darken the sun and leave a vast archaeological ruin in the event of our demise. Today, Earthly astronomers are looking for the darkness cast by alien engineering on this scale. With this level of technology we could even move stars around, albeit slowly. The simplest way would be to place a mirror on one side of the star to reflect some of its light into a beam, producing thrust in the opposite direction. Or the energy from a Dyson sphere could power ion engines to move the star a bit faster. We might use such a stellar engine to coast away from a predicted supernova, or to take two small dim stars and smash them together to make a brighter power plant.

If wrangling mere stars seems humdrum, how about harnessing the power of a supermassive black hole? We could catch the radiation from its accretion disk, or suck energy out of its spin. A spinning hole drags space-time around it in a region known as the ergosphere, which we could exploit in at least two ways. Roger Penrose of the University of Oxford has suggested using it to accelerate a stream of matter to high speeds, whereas University of Cambridge physicists Roger Blandford and Roman Znajek have devised a way to turn it into an electromagnetic dynamo. These could be the basis of a power plant a billion times as powerful as a Dyson sphere. It would have to be the size of our solar system, at least.

Even that might not be the limit of our ambition. Our

quest to understand the cosmos may lead us to build the ultimate particle accelerator, capable of reaching the immense energy where all forces are unified and the fundamental nature of space-time is finally revealed.

Brian Lacki at the Institute for Advanced Study in Princeton, New Jersey, has worked out some of the properties of this machine. One of them is immense size: to be able to boost particles to the required energies it would have to stretch at least a hundred times the distance from the sun to Pluto. This is just a lower limit, but making it even bigger should be possible. Such a long, thin object can be extended indefinitely without producing overwhelming gravitational stress, so we might be able to build a tower out to Orion and beyond.

At the limits of imaginable technology, we might even end up tinkering with the fate of the cosmos. If our hunger for power isn't satisfied by stars and supermassive black holes, then we might learn to create microscopic black holes and feed them with dust. This could unlock the mass-energy of inert matter, turning it into hot Hawking radiation that could be used to drive our interstellar industry.

According to calculations by S. Jay Olson of the University of Queensland in Australia, this could change the future of everything. With civilisation spreading through space at close to the speed of light, it would fill the cosmos with waste heat and so change its physical properties. The conversion of matter to radiation would even slow down the expansion of space a little, which puts our petty meddling with Earth into perspective.

Stephen Battersby

Sneezing season

? What causes hay fever, and why are only some people affected?

The responses that cause the symptoms of hay fever are nearly identical to the body's mechanism for destroying and expelling parasites. How these reactions come to be triggered by innocuous substances like pollen is not well understood. What is known is that the sensitisation process generally starts months or years before you notice any symptoms. The end result is a network of immune cells primed to react to specific allergens when you encounter them again in the future.

For some reason, more and more people are being sensitised. Allergic rhinitis affects people across the world but the highest recorded incidence among adults is 30 per cent in the UK, three times what it was in the 1970s. The US, Australia, New Zealand and other Western countries have experienced similar growth. The global average is about 16 per cent, meaning that more than a billion people have hay fever.

Some figures suggest it is more common among teenagers, and hay fever often first raises its head during the teenage years. But symptoms can suddenly start at any time of life, even in old age. Quite why it can take so long to come to a head isn't known. 'It could be due to changing hormone levels, or it may simply take time for sensitisation to develop,' says Stephen Durham, professor of allergy and respiratory medicine at Imperial College London. Drinking alcohol and smoking also increase the risk of developing hay fever and make symptoms worse. What's behind this epidemic?

Increased awareness could be part of it, but that's not the whole story – allergies of all kinds have seemingly become more common in recent decades.

The leading explanation is the hygiene hypothesis – the idea that decreased exposure to bacterial infections and parasites during early childhood disrupts the normal development of the immune system, causing it to pick fights with harmless substances. Growing up on a farm seems to protect children against hay fever and asthma, as does drinking unpasteurised milk.

But the hygiene hypothesis is not a complete explanation either. Certain countries such as Japan are extremely Westernised yet have very low rates of allergy – possibly because of underlying genetic factors that influence susceptibility. If you have one close family member with hay fever, you have a greater than 50 per cent chance of developing it yourself.

One recent twist is the suggestion that reduced contact with natural environments might lower the diversity of microorganisms living in and on us, and that this might tip the immune system towards an allergy-prone state. Ilkka Hanski at the University of Helsinki in Finland recently discovered that, compared with healthy individuals, people who are predisposed to developing allergies are more likely to live in built-up areas, and have less-diverse bacteria living on their skin.

One group of bacteria called *Acinetobacter* looks especially interesting because it seems to encourage immune cells to produce an anti-inflammatory substance called IL-10. 'Children growing up in homes that are surrounded by more forest and agricultural land tend to have more of these bacteria on their skin, and lower rates of allergy,' Hanski says.

Other possible factors that increase susceptibility include greater use of antibiotics during childhood, lower levels of vitamin D or exposure to certain chemicals. For now, nobody really knows.

Linda Geddes

Alternative history

? Neanderthals died out long ago, but their genes live on in modern humans. So how much of a Neanderthal are you?

In 2010, a group of geneticists announced that they had cracked the Neanderthal code. Using tiny fragments of 38,000-year-old DNA painstakingly pulled out of ancient bones, they had, against all odds, reconstructed a Neanderthal genome. We're used to hearing about ancient genomes these days, but at the time this was revolutionary. Until that point we only had bones to tell us about our extinct cousins, now we had the blueprint.

The geneticists decided to line up their Neanderthal genome against that of modern, living humans, and that revealed something amazing: there were bits that matched. Not just in the way that human DNA largely matches chimp DNA – when they looked closely, the team found that bits of modern human DNA were, in a sense, Neanderthal. The only explanation was that tens of thousands of years ago, a Neanderthal and a human had got down to it, had a hybrid child, and that child grafted fragments of Neanderthal DNA into the human family tree.

Today we know that 2 to 4 per cent of non-Africans' DNA is Neanderthal. The interbreeding happened after

Homo sapiens left Africa, so people who have a very pure African lineage don't have any Neanderthal DNA. And we know that people didn't all inherit the same bits – in total about 30 to 40 per cent of the Neanderthal genome is still knocking about, spread out higgledy-piggledy in millions of living humans. That DNA is not idle. So geneticists have been poring over those little bits to find out what traits some people might have inherited from the less fortunate species. If you have pale, freckly skin and red hair, there's a chance that the genes responsible were originally inherited from Neanderthals. You may also owe some of your immune system to them, and there's one gene people got from Neanderthals that governs the size of the tiny blind spot in our vision.

Of course, you could also flip the question on its head and ask how much of you is 'authentically human'. Which are the bits that we have and Neanderthals do not? There's a lot in there, of course, but one suite of genes is intriguing. They are found in our own cells and in dogs and cats, but not Neanderthal bones. They are interesting because they have previously been associated with domestication: dogs have them but not wolves, cats have them but not wild cats, and they are even found in a colony of Siberian foxes that have been domesticated, but not in their wild relatives. These genes give pets small skulls and narrower faces than their wild counterparts, and also make them less aggressive. So here's the lesson: many people are part Neanderthal. But what really sets us apart from them, and probably all our ancestors, is that we are the domesticated species.

Catherine Brahic

More human than us

? Is there is a continuum of consciousness from worms to birds to monkeys to humans? If so, what lies further along the line – what does it mean to be more conscious than a human?

It appeals to common sense that as animals become more complex they also get ever more consciously aware. For one thing, it's gratifying to have the continuum end with us. For another, it helps in justifying how we treat other animals. As Kurt Cobain put it, 'it's okay to eat fish, 'cause they don't have any feelings'.

But (sorry Kurt) common sense is often hopeless at explaining anything. Take consciousness itself. It's a word describing a phenomenon that everyone understands, but that is incredibly difficult to pin down scientifically. We've talked about it for millennia, and studied it for almost as long, but we still cannot say how or where in the brain the phenomenon of consciousness is generated.

One of the leading theories of consciousness, integrated information theory (IIT), sidesteps this problem by proposing that the experience of consciousness is constructed by data drawn from various locations in the brain. IIT says there is a measurement, phi, which represents the amount of emergent information possessed by a system.

If IIT is right, then it implies that all animals are to some extent conscious, and that there is indeed a continuum moving from simple to complex animals. But because we don't know where to look or how to measure it, it's hard to say where any of them are on the continuum. Or indeed, where we are.

We have no reason to think that human consciousness is the 'most' conscious a being can get. For a start, other animals have forms of consciousness without the same brains as us. Birds don't have a neocortex, the part of the brain we used to believe, with mammalian superiority, was essential for conscious thought. Instead they manage to do lots of complex cognitive processing – thinking – using a different part of the brain. I always remember encountering a group of jays in an aviary in Cambridge. The birds were habituated to the scientist I was with, but didn't know me. They hid behind a box and seemed to shout out, 'sorry, sorry'. It seemed as if they were apologising to their 'friend' for running away.

Most biologists who have worked with great apes will assure you they are conscious. One time a young orangutan made friends with me in the rainforest in Malaysian Borneo (it's a long story). The experience with him, and with the jays, left me in no doubt that I'd interacted with conscious beings. But it's not just my anecdotes and possible projections onto other animals – there is plenty of evidence of their conscious behaviour. You can even make the argument for plants. They are remarkably good at detecting and processing huge amounts of information, and acting on it. That's certainly something we'd call awareness, and maybe even consciousness.

So it's not really controversial to state that there's a continuum of consciousness. Each of us also experiences the points along the line. In our dreams we are still mostly ourselves, even if we happen to be flying or breathing underwater. If we're drunk or on drugs our consciousness operates differently. As IIT originator Giulio Tononi of the University of Wisconsin, Madison, puts it, consciousness was different when we were very young and will be different again when we are very old.

We're not at the peak of conscious possibility. What's next? Telepathy? Prophesy? Neuroscientist Anil Seth from the University of Sussex, UK, says the 'higher levels' of consciousness would not necessarily map to folk notions of cosmic togetherness – but hey, we can but hope. Possibly a higher consciousness would be far more attuned than we are to sensing another being's conscious state, and predicting its behaviour. To us this extreme empathy or hyper-intuition would look like mind-reading but would not require supernatural explanation, only a deeper understanding of neural processing.

Which brings us back to fish. How do we know they don't have any feelings? Since we don't know how pain and suffering are represented in our brains, let alone in other species, we can't say with authority what other animals are experiencing. It might be OK to eat fish, but not because we know they don't have any feelings.

Rowan Hooper

Hitting the snooze button

? Whether it's due to long days at work or long nights somewhere else – or both – it's a question many of us have asked: can I catch up on missed sleep at the weekend?

The need to sleep comes from a two-tier system. On one side there is sleep drive, also known as sleep pressure: the longer we are awake, the more this builds up, prompting our desire for sleep. This is the increasingly desperate need to sleep you might feel if you try to pull an all-nighter. As soon as you drop off, this sleep drive begins to dissipate.

Yet we don't feel more and more tired from the moment we wake up. For instance, people often feel more awake at eight in the evening than at four in the afternoon. That's because our sleep/wake patterns are also coordinated by the circadian clock in the brain. This produces what are known as alerting signals, which grow stronger for much of the day, to counter sleep drive. As bedtime approaches, these alerting signals drop off, and sleep drive overcomes us.

While we slumber, sleep drive diminishes until we can wake up feeling refreshed. Ideally, you shouldn't need an alarm clock to wake up in the morning – if you do, it's a sign you are accumulating a sleep debt. This will leave you feeling more and more tired as you fall into arrears. And the negative effects of getting too little sleep have been well documented – from memory loss to heart disease, weight gain and stroke.

So the simple solution is to repay this debt as soon as you can. In one study, researchers followed students who slept just four hours a night for six nights in a row. They developed insulin resistance (a precursor to type 2 diabetes), higher blood pressure and a rise in the stress hormone cortisol, as well as producing half the normal number of antibodies to this hormone. But all the effects were reversed when the students then caught up on the hours of sleep they had lost. Even so, not everyone is convinced that this kind of short-term recovery will alleviate the long-term health problems that result from regularly skipping sleep.

And there's another catch that might keep you awake at night. We know that shift work and jet lag, which play havoc with your body clock, also impair your health. Regularly sleeping at the wrong time can lead to diabetes, obesity and cancer, among other problems. And it seems that catching up on missed sleep at the weekend, a phenomenon known as social jet lag, can cause the same kinds of health problems as shift work.

Unfortunately, with work and school set to begin early in the day, those who suffer most are late chronotypes – people whose body clocks are naturally programmed to make them go to bed and get up late. The traditional view that these people are lazy is now looking more like health discrimination. The early bird catches the worm, whereas the night owl forced to get up early catches diabetes. The disparity is leading many to campaign for a change in school and office hours – in particular to protect the health of teenagers who tend to have a late chronotype. In the meantime, the healthiest solution is an early night for all.

Catherine de Lange

Encounters of the first kind

? What happens if we find aliens?

Thanks to the Kepler Space Telescope, we know the galaxy could hold as many as 30 billion planets similar to our own. The next generation of eyes in the sky, such as the James Webb Space Telescope, slated to launch in 2018, will search the atmospheres of such exoplanets for signs of life. Some think it's just a matter of time before we find out we're not alone. In April, NASA's chief scientist Ellen Stofan predicted we would have 'strong indications of life' on other planets by 2025. If she is right, how will we deal with the news?

What we detect will make a big difference to how we react, says Steven J. Dick, a former NASA historian and current astrobiology chair at the US Library of Congress in Washington, D.C. Any discovery that is less obvious than little green men landing during the World Cup final is likely to be met by years of questions and examination. Sara Seager, a planetary scientist at the Massachusetts Institute of Technology who is searching for another Earth, agrees. It will probably take time to confirm any initial findings, she says. 'There may not be an "aha" moment.'

A chemical imbalance in an exoplanet's atmosphere could be a sign of microbial activity. But an indirect result such as this will probably have only a short-term impact, says Dick. The apparent discovery of Martian nanofossils in meteorite ALH84001 in 1996 led to a media frenzy, and even US congressional hearings, before the furore died down in the face of increasing scepticism. Most now think that the meteorite does not hold the remnants of ancient alien life.

A decoded broadcast from intelligent aliens would be altogether different. Scientists and governments would have to assess whether the message was threatening and what, if anything, should be sent as a response. It would pose a challenge to certain religions too, says Dick. 'Does Jesus have to be a planet-hopping saviour to all ETs?' Some people might see intelligent aliens as saviours themselves, giving rise to new religions. Others may simply celebrate them as species that overcame their provincial squabbles to explore the universe.

In the longer term, even slight evidence of extraterrestrial life would spark a quest to understand the universal principles of biology, says Dick. We may find answers to questions such as: does life arise wherever the conditions are right, or is it a freak accident? Are there other types of genetic code? Does life always require carbon or water? Is Darwinian natural selection a universal, or are there other forms of evolution?

Perhaps most significantly, it would be the decisive blow

to the idea that humans are the centre of the cosmos or the reason for its existence. Instead, we would be forced to acknowledge our place as just one tiny branch of a vast galactic tree of life. 'I hope that people will find a new sense of peace and an understanding that we are not alone,' says Seager.

MacGregor Campbell

Cosmic coloniser

? Do humans have a responsibility to seed other planets with life? Or should we maintain a cosmic quarantine?

It's fun to speculate about aliens. But what if there are no aliens? It's been sixty-five years since Enrico Fermi first pointed out our solitude. Fermi estimated that it would take an advanced technological civilisation 10 million years or so to fill the galaxy with its spawn. Our galaxy is 10,000 times older than that. Where is everybody?

It's not as though we haven't been looking. Not for long, perhaps, and not very hard, but even a crude estimate suggests there should be other advanced civilisations capable of signalling over interstellar distances. And yet – nothing.

So what if we really are alone, or so isolated as to amount to the same thing? 'If we think we are the only life in the universe, we have a huge responsibility to spread life to the stars,' says Anders Sandberg of the University of Oxford's Future of Humanity Institute. 'If we are the only intelligence, we may have an almost equal responsibility to spread that, too.'

NASA astronomer David Grinspoon agrees, although he

hasn't given up on finding ET yet. 'We have these powers that no other species has had before,' he says. 'If we are it, if we are the best the universe has got, if we are the universe's sole repository of intelligence and wisdom and scientific insight and technology, it ups the ante quite a bit. We have a responsibility to preserve our civilisation.'

It won't be easy. First, we need to decide where 'to boldly go'. We don't know if humans can survive for any meaningful length of time anywhere except on the surface of Earth. 'Nowhere in our solar system offers an environment even as clement as the Antarctic or the top of Everest,' says the UK's Astronomer Royal Martin Rees. But some pioneers aim to give it a go anyway. Billionaire inventor Elon Musk is aiming to establish a self-sustaining colony on Mars in the next fifty years. 'By 2100, groups of pioneers may have established bases entirely independent from Earth,' Rees says.

Second, we are going to need some serious propulsion power but we don't yet know what that will look like. Third, we have to have some way to deal with interstellar dust, which could create catastrophic collisions with our craft at the speeds we would need to attain. Fourth, we would need some kind of artificial gravity onboard, otherwise the crew will suffer massive, possibly fatal, health issues.

There will undoubtedly be many more obstacles that we have yet to confront or even imagine. But Sandberg and others are optimistic we can overcome them. Even if we can't, we could play the longer game and attempt to seed the galaxy with life via 'directed panspermia'. The basic idea is to launch microorganisms into space in the hope that they will crash-land on a planet or moon suitable for life, and eventually evolve into a self-aware, intelligent species.

Science fiction author Charles Stross suggests isolating spore-forming *Archaea* and photosynthetic bacteria that can survive for long periods in the harshest of environments. 'Put them on rockets and fire them out of the solar system,' he says. 'Almost all will perish, but if you launch a hundred tonnes of spores every year for a century, maybe sooner or later something will work.'

There would be no real payback for us, except perhaps returning a favour. Life on Earth may have started by directed panspermia. If so, our ultimate purpose may be to pass it on again, like a chain letter through the cosmos.

Michael Brooks

The trouble with tumours

? Why haven't we found a cure for cancer yet?

'If we can send a person to the moon, why can't we cure cancer?' It's a familiar grumble, but it speaks volumes about how most of us view cancer. Surely it's just a case of unwanted lumps – cut 'em out and problem solved! Often that does work, provided the lump is still small, accessible and that it is the only lump.

And therein lies the rub. Hardly anyone dies from the initial, cancerous lump, called the primary. What kills people are the secondaries, fragments of the original lump that break off, spread to other parts of the body and start growing as new lumps, often in organs distant from the primary. Once this process, called metastasis, gets going, cutting out the primary does no good, because the secondary lumps will carry on growing and invading other parts of the body.

Often the secondaries spread without a person even knowing that they have a primary. By the time they feel symptoms, maybe because a lump becomes visible or starts pressing on nerves and causing pain, the secondaries are too numerous to treat through surgery.

But we do have some natural defences. Our body's immune system works primarily to identify and kill invaders, like bacteria and viruses, but as part of its daily housekeeping operations, it can also spot and destroy abnormal cells – the kind that can lead to cancer. Likewise, our cells have their own internal quality checks that trigger cell suicide if anything is dangerously wrong.

Abnormal cells, lumps and even tiny cancers probably grow in us all frequently, but our immune systems are fantastic at spotting and destroying them. Rarely, however, a tiny tumour can evade detection. Researchers are learning that tumours can evolve 'don't kill me' signals that dupe the immune system into inaction. Tumours may also develop as part of a faulty attempt to heal a wound or grow tissue for repair. This may fool the immune system into thinking that the process under way is a healing one rather than a lethal one.

Many of the genes and processes driving the growth of tumours are otherwise only active in foetuses. It's as if the tumour is desperately trying to grow new body parts at random within yours. And the speed at which a foetus develops might help to explain why some cancers grow and spread so fast, making them difficult to treat.

When trying to develop treatments, the other major complication is that cancers originate from random cells in random organs in random people, so they're all completely different. And cancer cells are a mess, both genetically and in the way they behave. They cut loose from their normal job and go haywire, switching on genes that should be off and activating forbidden cellular processes, enabling them to multiply uncontrollably or burrow out of one organ and spread to another. Worse still, they become a jumble of different cells when they spread, and even within a single tumour. It's less an abnormal cell or tissue you're trying to kill, more a constantly evolving ecosystem.

Thankfully, medicine is making gradual progress. The best solution once a cancer has been diagnosed is to cut it out surgically. Usually, patients also receive chemotherapeutic drugs that – unlike the surgeon's knife – can circulate throughout the body and pick off any secondaries before they become a problem. Most traditional chemotherapy drugs kill cells that are actively multiplying. This works because cancer cells multiply rapidly, unlike most other adult cells. However, there are a number of places in the body where healthy cells routinely divide – such as the lining of the gut and mouth, not to mention the hair follicles – and these get caught in the crossfire. This is what causes the severe side effects many patients endure. To make matters worse, tumours contain so many messed-up, mutated cells to start with, there may be a handful that are resistant to the chemotherapy. These can survive the treatment and regrow, evolving within the patient.

A different tactic involves finding genetic markers in cancer cells that single them out as abnormal, and which can be targeted by medication. Drugs like trastuzumab (sold as Herceptin) for breast cancer work like this, and have dramatically increased survival in patients who have the correct marker for the drug to target. While optimism grew over the past decade, there soon came the realisation that some cancers could evolve resistance to these medications too, by shedding their targeted site, rendering them invisible to the drug.

More recently, researchers have been returning to a solution that aims to prime the immune system to fight the cancer. The beauty of this is that the immune system can evolve too, and so potentially keep pace with a cancer doing the same thing. These new drugs like nivolumab and pembrolizumab work by tearing up the deceptive cloak of the tumour and allowing the immune system to do the rest. Most importantly, this treatment can keep tabs on and destroy the cancer even if it evolves, something that no single silver-bullet drug can do. These drugs have had profound impacts on the survival of patients with otherwise fatal skin cancers and lung cancer – the biggest cancer killer worldwide – helping some live for years rather than just a few months.

So the future is looking bright. There may never be a single 'cure for cancer'. But re-engaging the immune system shows great hope – it could be the first solution that adapts and evolves within each patient to keep their unique form of cancer under control. It will be a long and difficult trip, but this very Earthly moonshot is taking off.

Andy Coghlan

Icy slide

? Why is ice so slippery?

For physicists no less than figure skaters, ice is remarkably hard to get a grip on. The overwhelming consensus is that ice has low friction because of a thin film of liquid water coating its surface. Hence skaters balanced on thin metal blades can glide smoothly across the ice rink, but grind to a halt on the wooden floor beyond. The tricky part is how this liquid layer forms. More than a century of research has brought us little closer to a definitive answer.

It all started in June 1850, when Michael Faraday told an audience at London's Royal Institution of how pressing two ice cubes together led to them forming a single block. He attributed this to the appearance of an intervening film of water that quickly refreezes. For many years, the appearance of this layer of water was put down to pressure. In fact, even a person of above-average weight on a single skate generates far too little pressure to account for the observed melting, says Anne-Marie Kietzig of McGill University in Montreal, Canada: 'The mathematics doesn't work out.'

Instead, Kietzig argues that the main player is frictional heating. The movement of a blade across the ice, for instance, easily generates enough heat to melt some of it.

You might think that would be the end of it. But Changqing Sun of Nanyang Technical University in Singapore has other ideas. He argues that since ice is slippery even when you're standing still, friction cannot be the whole story. 'Mechanisms such as friction heating and pressure melting have been ruled out,' he says.

According to Sun, the assumption that the slippery layer coating ice is a liquid is also fundamentally flawed. He says this layer should properly be called a 'supersolid skin' because the weak bonds between H_2O molecules at the surface are stretched, but unlike in liquid water none of them are broken. He also argues that this elongation of bonds ultimately produces a repulsive electrostatic force between the surface layer and anything it comes into contact with.

He compares the effect to the electromagnetic force that levitates Maglev trains, or the air pressure a hovercraft generates beneath its hull. If he's right, his model helps to explain many of the layer's properties, including its remarkably low friction. 'I believe the problem has been completely resolved,' says Sun.

Most in the ice field are not convinced. Gen Sazaki at

Hokkaido University in Sapporo, Japan, who made the first direct observations of this layer in 2013, prefers to call it a quasi-liquid. He thinks it represents a transitional stage between solid and liquid as the temperature goes up.

For Sazaki, understanding how this mysterious sheet of H_2O forms is still some way off. Even when it comes to something as familiar as slipping on ice, he says, 'reality is much more complicated than we expected'.

Gilead Amit

Future direction

[?] Most of us have wished, at one time or another, that we could turn back the clock. Why do we only move forwards in time?

There is a reason we say time goes by: it seems to flow. No matter how still we stand in space, we move inexorably through time, dragged as if in a current. As we do, events steadily pass from the future, via the present, to the past.

Isaac Newton saw this as a fundamental truth. 'All motions may be accelerated and retarded, but the flowing of absolute time is not liable to any change,' he wrote.

So how does time flow, and why always in the same direction? Many physicists will tell you that's a silly question. 'The idea that time can in some meaningful sense be said to flow, it's just a complete non-starter,' says Huw Price, a philosopher at the University of Cambridge.

For time to flow, it must do so at some speed. But speed is measured as a change over time. So how fast does time flow? George Ellis, a cosmologist from the University of Cape Town, South Africa, has an answer: 'One second per second.' Price says that's meaningless. Even if time were standing still, it could be said that for every second that passes, one second passes. Indeed, if that's a measure of flow, we could say that space flows: it passes at one metre per metre.

Ellis is up against one of the most successful theories in physics: special relativity. It revealed that there's no such thing as objective simultaneity. Although you might have seen three things happen in a particular order – A, then B, then C – someone moving at a different velocity could have seen it a different way – C, then B, then A. In other words, without simultaneity there is no way of specifying what things happened 'now'. And if not 'now', what is moving through time?

Rescuing an objective 'now' is a daunting task. But Lee Smolin of the Perimeter Institute for Theoretical Physics in Waterloo, Canada, has given it a go by tweaking relativity. He argues that we can rewrite physics in a way that includes 'now' if we sacrifice some of our objective notions of space.

Most physicists aren't having it. The general consensus is that time is more or less just like space – an immutable dimension, stretched out through a four-dimensional 'block universe'.

'Every moment in that universe has a past, present and future,' says Sean Carroll from the California Institute of Technology in Pasadena. 'A person is described as a history of moments, and those moments all have a feeling that they're moving from the past to the future.'

That doesn't answer the question so much as shift it. If time does not flow, what makes us think it does?

Michael Slezak

Explain that pane

[?] Is glass really a liquid?

Forget the hoary myths peddled by tour guides in old European churches and cathedrals. Medieval windowpanes are sometimes thicker at the bottom not because of the slow flow of glass over centuries, but because of the uneven way molten glass was originally rolled into sheets in the Middle Ages.

Glass is not a slow-moving liquid. It is a solid, albeit an odd one. It is called an amorphous solid because it lacks the ordered molecular structure of true solids, and yet its irregular structure is too rigid for it to qualify as a liquid. In fact, it would take a billion years for just a few of the atoms in a pane of glass to shift at all.

However, not everything about glass is quite so clear. How it achieves the switch from liquid to amorphous solid, for one thing, has remained stubbornly opaque. When most materials go through this transition between liquid and solid states, their molecules instantly rearrange. In a liquid the molecules are moving around freely, then snap! – they are more or less locked into a tightly knit pattern.

But the transition from the glassblower's red-hot liquid to the transparent solid we drink from and peer through doesn't work like that. Instead of a sudden change, the movement of molecules gradually slows as the temperature drops, retaining all the structural disorder of a liquid but acquiring the distinctive physical properties of a solid. In other words, in all forms of glass we see something unusual: the chaotic molecular arrangement of a liquid locked in place.

The process underlying this strange behaviour remains an

open question. 'The number of explanations almost matches the number of researchers,' says Hajime Tanaka at the University of Tokyo in Japan. One possibility is that it's all down to energy use. According to the laws of thermodynamics, which govern how energy is transferred within a system, every collection of molecules is driven to find an arrangement with the lowest possible energy. But within any given system some patches do better than others, meaning different groups of molecules settle into different configurations – and, overall, into an irreconcilably chaotic arrangement.

But even if we put it down to thermodynamic laws, it's not clear what exactly drives the strange behaviour of glass. The push for low energy might be the prime mover. Then again, it could be the irrepressible tendency towards a maximum state of disorder. That's a perfectly plausible proposal, though it raises the troubling question of how ordered solids manage to survive.

Tanaka is not giving up just yet. 'So far crystallisation and glass transition have been studied independently,' he says. But Tanaka believes that glass may form in a manner not all that different from crystals, which have proved an easy target for analysis thanks to their repeating geometric structures. If he's right, maybe glass will finally become crystal clear.

Gilead Amit

Paltry pull

 Why isn't an entire planet's gravity enough to rip a magnet off your fridge door?

Hang on, you might say: gravity is strong enough to keep my feet on the ground, and no space agency firing craft into orbit would ever describe gravity as weak. But the mystery for physicists is why that force is so puny compared with the electromagnetic force that it doesn't rip that magnet off your refrigerator – we're talking about the pull of an entire planet, after all.

This mismatch between gravity's strength and that of the other forces of nature goes by the name of the hierarchy problem. Because, uniquely, gravity is not yet described by a quantum theory, it's not easy to quantify the problem's size, but one measure is the Planck mass, a quantity that gets bigger the weaker gravity is. In our cosmos the Planck mass is huge. It is some 10 quadrillion times bigger than the mass of the W and Z bosons that define the strength of the weak nuclear force, for example. In fact, it is huge compared with all masses that pop up in the standard model. 'The question is not why the Planck mass is big; the question is why it is big compared to the masses of all the known particles,' says theorist Matt Strassler of Harvard University. 'The puzzle is something you can phrase either as the Planck mass being large or particle masses being small.'

Explanations following the first route often invoke the idea of 'fine-tuning': that we just happen to live in an unnatural part of the universe where gravity is just right, so atoms, stars, planets and people have come to exist. Or they propose large extra dimensions of space into which gravity 'leaks', so it appears diluted to us.

Alternatively, we can focus on the Higgs field, which generates particle masses. The low mass of the Higgs boson, discovered in 2012, indicates this field is not particularly strong, keeping all particle masses on the low side. Theories such as supersymmetry and technicolor focus on as-yet-undiscovered particles or forces whose effect is to restrain the Higgs field to the observed strength of almost – but not quite – zero.

Experiments aren't helping decide between these options as yet. Supersymmetry – or indeed anything new besides the Higgs boson – has so far failed to make its presence felt in the particle smashes going on at the Large Hadron Collider (LHC). 'Not finding anything else yet leaves us at sea,' says Strassler.

The hope is that future runs of the LHC, now operating at maximum energy and generating more particle collisions than ever, could give us more of a clue. That's why the recent appearance of blips in LHC data, indicating the existence of a particle six times as massive as the Higgs boson and not predicted by the standard model, has made many a physicist's heartbeat swifter. But it is still too early to say whether these blips will persist – or what, if any, solution to the problem of gravity they might support.

Michael Brooks

Not before your time

? How young can you die of old age?

Loosely, someone dies of old age when ageing is responsible for the changes that have brought their body to terminal failure. However, deciding whether a particular person's exit can be classified as dying of old age will be tricky until we've managed to develop more accurate ways to measure the amount of damage done by ageing.

As you grow older, you become increasingly likely to develop and to die from many conditions, including cancer, heart disease, neurodegenerative disorders and osteoporosis. While lifestyle and genetics can make some of these illnesses more likely to strike, a good lifestyle and a lucky genetic hand will not save you – even people who don't smoke, take daily exercise and eat healthily get old and die.

That's because there's no escape from the processes of ageing; we lose the ability to repair and regenerate tissue with stem cells, and we accumulate DNA damage as well as old and dying cells. These processes start very early – perhaps even while in the womb – but the link between age and death doesn't really get started until puberty. From around the age of thirteen or so, a person's chance of dying climbs exponentially with every further year.

But classifying a teenage cancer death as dying of old age doesn't feel right, even if DNA damage played a part. So when does someone become old? It's a sliding scale, and experts joke that a general rule is that it's fifteen years from whatever your age happens to be right now.

But there are some more practical definitions too. One of these is multimorbidity. From around the age of seventy, we start to accumulate age-related chronic illnesses, like heart disease, arthritis, dementia and diabetes. When the Newcastle 85+ Study began in 2006, they found that, out of more than 1,000 eighty-five-year-olds, three-quarters had four or more medical conditions.

So in some respects, dying of a stroke in your sixties isn't necessarily death from old age, unless you also happen to have a few other serious disorders. When a person has multiple diseases, the cause of their death can be less clear-cut, and dying of old age starts to become a more useful term.

A defined, immediate cause of death is more likely to be listed on a person's death certificate – for example, cardiac arrest or pneumonia. But one definition of dying of old age could be that, if it hadn't been one cause today, then it might have been a different one tomorrow.

Everyone ages at a different rate, and it isn't possible to

say exactly how young a person could die from old age. But just as demography can help us determine average life expectancies (about eighty-four and eighty-six for male and female sixty-five-year-olds living in the UK today), statistical methods can help us work out how old you have to be before death from old age starts to become increasingly likely.

In this way, David Melzer at the University of Exeter, UK, and his colleagues can predict death due to ageing for particular demographic cohorts. Using statistical models, they can classify lives as long, intermediate, short, or prematurely ended. Doing this using data from people born in 1951, they find that deaths before the age of sixty-three in women, and fifty-eight in men, are statistically premature, giving a guide for when people of each gender are likely to start dying due to age. But there's a fair amount of variation, and their calculations suggest that the majority of deaths from old age will occur after the ages of eighty-one in women, and seventy-eight in men. So we can say that dying in your eighties can't yet be considered 'before your time' – although we suspect many octogenarians would disagree.

Penny Sarchet

Caught in the moment

? How long is now?

How long is a piece of string? The question 'how long is now?' has a similarly unanswerable, almost metaphysical feel. But science can answer it; we just end up with many different answers, from 'zero length' to 'as long as you like', via all points in between. Just like the string, then.

The feeling of existing permanently in a present moment

as time flows past us is one of the most essential of human experiences. The 'now' it defines is a point in time that has no duration, just a personal portal between past and future. But that sort of now is too abstract even for physicists. Time does not flow – it is a dimension like space that just *is*, and all of it exists for all time.

This 'block universe' thinking is made necessary not least by Einstein's relativity, which says that now is, well, relative. Take a photon from the cosmic microwave background. It reveals to us a picture from just after the big bang 13.8 billion years ago. But from the photon's perspective, its journey from back then takes no time at all: our now, the now of the big bang and all the nows in between happen simultaneously. From a cosmic perspective you can define now to be as long as you like – your answer will be equally right (or wrong).

But if the present is just an individual illusion created in our brains, we can still ask how long *that* now is. Marc Wittmann of the Institute for Frontier Areas of Psychology and Mental Health in Freiburg, Germany, is author of the book *Felt Time: The Psychology of How We Perceive Time*. He thinks there are broadly three answers.

First comes the 'functional moment'. This is the time we need between two stimuli – a click in the right and left ears, say – to distinguish their order. It varies between senses, but is generally a few tens of milliseconds.

Our brains need a little longer to stitch various stimuli together into a conscious 'experienced moment'. Various experiments show that this lasts between two and three seconds. Listen to a metronome, for example, and you will have no trouble tapping your finger or foot to its beat provided the next comes within this time frame. Any longer, however, and you are likely to hear yourself counting under your breath to synchronise your beats.

Our individual experienced moments are then further stitched together by our working memories to produce a third now – the sense of continuity and time passing. The length of this now is tens of seconds, although it varies considerably depending on the density of stimuli our brains are working with. 'The feeling that time drags – you're waiting for the bus, you're cold, your smartphone isn't working – that's very real,' says Wittmann. Conversely, our brains ratchet up the processing speed if we're exploring new places or having new experiences.

This is the central truth of now, as far as there is any. 'Time and self are intricately linked,' says Wittmann. So perhaps the only valid answer is to live in the now – however long that is.

Richard Webb

Acknowledgements

'The Last Word' owes its success to a great many people, not least the combined efforts of Mick O'Hare, Jeremy Webb, the *New Scientist* subs desk and art team, Beverley de Valmency, Georgina Laycock and the team at John Murray, and all of our wonderful contributors, including Jon Richfield, Mike Follows, David Muir, Eric Kvaalen, Lewis O'Shaughnessy and many more besides.

Index

accidents: caused by stooping
    82–3
ageing 284–6
aggregate pellets 239–40
aircraft:
    condensation trails 205–6
    flying rules 197–8
    fog disruption 186–7, 192
    military drone design
        202–3
    piston engine noise 203–5
    temperature 178–80
    windows for pilots 191–4
    wing shape 184–6
albinos 25–6
alcohol:
    attention span 243
    microbial protection 84
algae:
    Earth colonisation 117–19
    pond 'pea-soup' clearance
        119–21
aliens: discovery consequences
    270–2
aluminium: outdoor furniture
    70–2
anemometers 195
animals: see also specific
        animals
    consciousness 266–8
    domestication 265
    fear (or not) of humans
        7–9
    fruit eating 112–13

herd protection 12
moss animals 238–9
navigation 152
pack hunting 14–17
self-medication 17–19
sleep position 30–1
ultrasound deterrents
    215–17
aphids 68
archery: wood choice for
        longbows 130
asteroids:
    discovery 248
    shape 44
attention span 241–3
avian flu management 123

babies: crying on film 106–7
bacteria:
    genetic resistance 92–4
    hygiene hypothesis 263
    rice food poisoning 58–9
badgers: wasp nest raids 14
ballast for railway sleepers
    182–4
baths: leftover house heating
    175–6
bats: perching 29–30
batteries: exploding 223
bees:
    fear (or not) of humans
        7–9
    honey cultivation 9–10
bicycle staying upright 245–6

birds:
  avian flu 123
  clucking hens 28
  colour aberration 25–7
  dinosaur ancestry 32–4
  pack hunting 14–17
  perching 29–30
  sunbathing 26–8
black holes:
  colliding 49–50
  future energy generation
    260
blackbirds: colour aberration
    25–7
blood:
  circulation systems 73–4
  diseases 93
  hypertension 108
bobsleigh tracks: ice stickiness
    180–1
body image: changing 82–3
boiling point:
  microwave drinks reheating
    54–6
  pan lid 56–8
brain:
  attention span 241–3
  consciousness definition
    250–1, 266–8
  experience of now 287–8
  inner speech 257–9
  thought location 91
bubbles:
  boiling point 54–6
  colour 169–70
  dropped beer bottle 60–1
  milk foam 61–2
  mushroom shape 160–1
burping:
  gut noises 92
  weight (mass) effect 100
butterflies:
  fear (or not) of humans 7–9
  flight pattern 130–1
  insect or not 10–11

camouflage: stripes 4
canals: tidal effect 47–8
cancer: cure discovery 274–7
candles: melting 64
cars: radio aerials 219
caterpillars: number of legs
    10–11
cats:
  domestication 265
  eye pupil shape 20–2
  ultrasound deterrents
    215–17
chlorophyll 108–10
Christmas tree smell 134–5
circadian clock 269
cities: floating at sea 228
clay: aggregate 239–40
clocks: 'polar clock' 37–8
clouds:
  aircraft condensation trails
    205–6
  anti-crepuscular rays
    136–7
  sea phytoplankton effect
    116–17
  sunburn 137–9
colour:
  distance effects 156–7
  soap bubbles 169–70
  subatomic particles 147–9
colour-blindness: lions 4
comet shape 44
common cold vaccine 124–5
concrete:
  railway sleeper beds 183,
    184
  road surfaces 189–90
consciousness: definition
    250–1, 266–8
cotton clothing 221–2
crabs: sand balls 235–6
crows: pack hunting 16
cuttlefish: eye pupil shape 21
cycling:
  brakes 215

cycling (*cont.*):
  keeping a bicycle upright 245–6
  temperature at speed 178–9
  tyre tread depth 201

death: old age 284–6
dental enamel 89–90
dialect 255–7
dinosaurs:
  DNA from mosquito stomachs 34–6
  what if they had survived? 32–4
disease resistance 85–6, 92–4
dishwashers: glasses cracking 66–7
dodo: extinction cause 8
dogs:
  domestication 265
  nose colour 17–19
  self-medication 17–19
drinks:
  alcohol microbial protection 84
  beer bottle dropping 60–1
  gold flake vodka 65–6
  leaking mugs 52–3
  microwave reheating 54–6
  milk foaming 61–2
  tooth decay 89
  wave patterns 173–5
drones: design requirements 202–3
drug resistance 92–4
dung beetles 22–5

eagles: pack hunting 16
Earth:
  algae colonisation 117–19
  first life 254–5
  geothermal energy 144
  magnetic poles flipping 150–2
  mountain size 44
  pressure at core 149–50
  sea phytoplankton climate regulation 116–17
  shape measurement 46
  weight difference between equator and poles 142–3
eggs:
  scrambling 59–60
  shell formation 232–3, 236–7
electricity:
  demand peak management 225–6
  geothermal energy 144
  phone charger function 225
  wind turbine speeds 214–15
elephants: sleep position 30–1
empathy 268
energy generation: *see also* electricity
  future galaxy power 259–61
enzymes: chlorophyll genetic engineering 109–10
evolution:
  alien discovery consequences 271–2
  first life on Earth 254–5
  Neanderthal ancestors 105–6, 264–5
exoplanets 45–6, 249
eyesight:
  cats' pupil shape 20–2
  colour perception at distance 156–7
  herd protection 12
  light compensation 79–80
  lion colour-blindness 4
  patterns seen when eyes closed 74–5
  relative importance 78–9
  underwater 4–5

fabric:
    cotton vs synthetic 221–2
    spacesuits 229–30
    wool pilling 207–8
faces: archaeological recon-
    struction 104–5
faeces: personal odour 102
fasciation 133–4
films: crying babies 106–7
fingerprints 86–8
fire: chemical reaction 42–3
fish:
    aversion due to odour 94–5
    sea odour 116–17
flag flapping 172–3
flatulence:
    gut noises 92
    weight (mass) effect 99–100
flies:
    odour attraction 6–7
    swatting failure 7–8
flight: see also aircraft
    birds vs bats 29–30
    gut bloating 190
    tower avoidance 197–8
flowers: multiple heads 133–4
foaming:
    milk 61–2
    seawater 234
fog:
    airport disruption 186–7,
        192
    ice stickiness 180–1
food:
    aversion due to odour 94–5
    meat eating 96
    rice poisoning 58–9
    taste and smell senses
        78–9
    tooth decay 89
    UK population supported
        without imports 131–3
fruit flies: odour attraction 6–7
fruit: seed distribution by
    animals 112–13

furniture: metal outdoor
    70–2
galaxies:
    colonisation 272–4
    light emission 177–8
    spiral arms 50–1
    technology to change our
        galaxy 259–61
gamma ray colour 148
geckos: eye pupil shape 21
genes:
    chlorophyll genetic
        engineering 108–10
    Neanderthal ancestors
        105–6, 264–5
    pathogen resistance 92–4,
        107–8
geothermal energy: electricity
    generation 144
giraffes: sleep position 30
glass:
    cracking 66–7
    liquid or solid? 281–2
goats: odour 11–12
goldfish: attention span 242
golf tee behaviour 164–6
goosegrass: animal medication
    17–19
gravity:
    Earth's core 149–50
    strength vs electromagnetic
        force 282–4
gut:
    bloating when flying 190
    noises 92

hand drying: energy use 224
Harris's hawk: pack hunting
    16–17
hay fever 262–4
head:
    facial reconstruction 104–5
    thought location 91
    tilting 80–2

hearing: *see also* noise
    change comparison 79–80
    location perception 227–8
    relative importance 78–9
hearing aids: TV and radio
    signal reception 220
hearts: blood circulation and
    gas exchange 73–4
heat: air friction 178–80
Heathrow: temperature 146–7
Helmholtz resonator 200–1
hens: clucking 28
heraldry: colour perception at
    distance 156
Higgs boson 283–4
horseflies: stripes as repellent
    3–4
humans: *see also* specific
    body parts
    consciousness 266–8
    moon's effect on weight
    46
    Neanderthal ancestors
    105–6, 264–5
    particle beam effects 176–7
hyperacusis 77–8
hypertension: salt and genetic
    variations 108

ice:
    fog stickiness 180–1
    slipperiness 277–9
immune system:
    cancer cure potential 275,
    277
    hay fever 262–4
inner speech 257–9
insects:
    definition 10–11
    fear (or not) of humans 7–9
    stripes as repellent 3–4
International Space Station:
    landing on the moon
    48–9
internet: data volume 218

jelly balls 238–9
Jurassic Park 34–6

kettles: limescale prevention
    68–9
knife sharpening 62
Kuiper belt 248, 250

ladybirds: fear (or not) of
    humans 7–9
lamp post wobbling 170–3
language:
    early development 100–1
    English dominance 255–7
laundry: drying stiff 69–70
legs: caterpillars 10–11
light:
    colour of subatomic
    particles 147–9
    dissipation 177–8
    eye pupil shape 20–2
    luminous watch dials 161–4
    photography polarising
    filters 166–7
    polarisation and stripes 3–4
    polarisation and time of
    day 37–8
    rainbow spectrum range
    139–41
    refractive indices 4–5,
    166–7
    vitamin D requirements
    97–8
limescale prevention 68–9
lions: colour-blindness 4
longbows: wood characteris-
    tics 130
lugworms 235

magnetic poles flipping 150–2
magnets:
    Earth's gravity effect 282–4
    field lines 157–8
    packaging for bulk trans-
    port 159

malaria:
  drug resistance 93–4
  sickle-cell anaemia 107–8
mammals: aquatic eyesight 4–5
maps: distortion of countries
    144–6
measurements: 16 inch
    intervals 72
meat eating 96
memory:
  inner speech 258–9
  Shakespeare's complete
    works 84–5
metal:
  curling 68
  floating gold flakes 65–6
  outdoor furniture 70–2
  photography polarising
    filters 166–7
microphone feedback 211–12
microwaves: drink heating
    54–5
midges: bite reactions 32
milk: foaming 61–2
molehill location 22–3
moon:
  aspect facing Earth 38–41
  definition 249
  effect on human weight 46
  exploration 252–4
  International Space Station
    landing 48–9
  tidal effect 41–2, 47–8
mosquitoes: dinosaur blood
    34–6
moss animals 238–9
mountains:
  gravitational effect 46
  maximum size 44
MP3 players: weight (mass)
    209–10
mugs: leaking 52–3

Neanderthal ancestors 105–6,
    264–5

noise:
  aircraft piston engines
    203–5
  microphone feedback
    211–12
  open car windows 200–1
  pain from loud noises 76–8
  road surfaces 188–90
  wind noise on video 198–9
nose-picking 103
noses: dogs 17–20
now 286–8
numbers: word evolution 100–1

odour:
  aversion to fish 94–5
  Christmas tree smell 134–5
  dog nose ability 18–19
  faeces 102
  fruit fly attraction 6–7
  goats 11–12
  sea 116–17
oil spills: dispersants 121–2
ozone: sky colour 139–41

paper: curling 67–8
paraffin: recrystallisation 64
pelicans: pack hunting 15
phones:
  charger function 225
  predictive text 231
phosphenes 74–5
photography polarising filters
    166–7
photoperiodism 113–14
photosynthesis:
  genetic engineering 108–10
  pond 'pea-soup' 121
phytoplankton: sea odour
    116–17
planets:
  colonisation 272–4
  definition 247–50
  detection of other systems
    45–6

planets (*cont.*):
    maximum mountain size
      44
plants:
    clay aggregate growing
      medium 240
    climbing direction 125–6
    consciousness 267
    daylight length effects
      113–14
    multiple flower heads
      133–4
    sacrificial planting 68
    thorns 112–13
Pluto 247–50
polar bears: food preferences
    12–13
polar clock 37–8
polyester clothing 221–2
positivity bias 258–9
Post-it notes 154–6
pottery: self-glazing pots
    168–9
praying mantis: fear (or not)
    of humans 7–9
predictive text 231
pregnancy: body image
    awareness 83
pressure:
    aircraft flight 190
    boiling point 56–8
    dropped beer bottle
      60–1
    Earth's core 149–50
    plumbing systems 63
proprioception 82–3

radio aerials on cars 219
radio waves: time telling on
    cloudy days 38
railtracks:
    electrical power loss in rain
      187–8
    sleeper ballast 182–4
    vibrations 217

rainbows: spectrum range
    139–41
raspberries: hairs 111
razor sharpening 62
relativity 279–80, 287–8
remote controls 219
resonance 200–1
rice: food poisoning 58–9
road surfaces: noise 188–90

salt: hypertension and genes
    108
sand: balls 234–6
scrotum: wrinkliness 98–9
sea: *see also* odour
    foam 234
seals: breeding grounds 12–13
seeds:
    animal distribution 112–13
    daylight length effects 114
senses:
    deprivation 91
    ranking 78–9
    relative comparison 79–80
Severn bore tide 47–8
Shakespeare's complete works
    84–5
sheep: sterility in Australia 99
shoe laces: tightness 79–80
sight *see* eyesight
similects 256–7
skin:
    bite reactions 31–2
    fingerprints 86–8
skulls: facial reconstruction
    104–5
sky: ozone effect on colour
    141
sleep:
    birds 29
    catching up on missed
      sleep 268–70
    position 30–1
    sneezing 75–6
slug trails 237–8

smell *see* odour
smell (sense): relative import-
     ance 78–9
sneezing:
     hay fever 262–4
     whilst asleep 75–6
sound: *see also* noise
     dissipation 177–8
     location perception 227–8
     relative comparison 79–80
     wind noise on video 198–9
space exploration 252–4,
     273–4
spacesuits 229–30
spiders:
     biting 31–2
     ultrasound deterrents
          215–17
Star Trek test 247, 249
stars: *see also* sun
     birth 51
     brown dwarfs 250
     detection of planets 45–6
     future energy generation
          259–60
     light emission 177–8
stomach noises 92
stone: ballast for railway
     sleepers 182–4
stooping: causing accidents
     82–3
stripes: evolutionary function
     3–4
strokes: head tilting 80–2
subatomic particles:
     colour 147–9
     particle beam effect on
          humans 176–7
sugar: moisture and hardness
     53–4
sun:
     anti-crepuscular rays 136–7
     effect of dousing with
          water 42–3
     galaxy orbit 51

light emission 148
tidal effect 42
time telling on cloudy days
     37–8
vitamin D requirements
     97–8, 141–2
sunburn 137–9
supernovae 43, 51
supersymmetry 283–4
surfing wave size 153

tape measures: imperial
     intervals 72
taste: relative importance
     78–9
teeth 89–90
temperature:
     air friction 178–80
     effect on cracks 52–3
     leftover bath water heat
          175–6
     UK levels 146–7
texting 231
thinking: location in head 91
thorns on plants 112–13
tides:
     moon effect 41–2, 47–8
     surf wave size 153
time:
     length of now 286–8
     relativity and flow 279–80
     sun position on cloudy
          days 37–8
toilets: water moving 63
touch: relative importance
     78–9
towels: drying stiff 69–70
trains: station announcement
     signs 212–13
trams vs trolleybuses 195–6
trees: *see also* wood
     Christmas tree smell 134–5
     daylight length effects
          113–14
     pollution resistance 115

tsetse flies: stripes as repellent
3–4
tuberculosis resistance 85–6
tunnel ventilation 195
tyres:
    road surface noise 188–90
    tread depth 201–2

ultrasound: animal deterrent
    use 215–17

van der Waals' forces: Post-it
    notes 154–6
ventilation on underground
    railways 195
viruses:
    avian flu management 123
    common cold vaccine
    124–5
vitamin D:
    biosynthesis in birds 27–8
    sunlight exposure 97–8,
    141–2
vodka: floating gold flakes
    65–6
vortices:
    lamp post wobbling 170–3
    open car windows 200–1

wasp nests: badger raids 14
watches: luminous dials 161–4
water:
    boiling in covered pan
    56–8
    boiling-point bubbles 54–6
    bubbles' mushroom shape
    161
    clay aggregate treatment
    240
    dousing the sun 42–3
    ice slipperiness 277–9
    ice stickiness 180–1
    leftover bath water heat
    175–6
    moving in toilet 63

pond 'pea-soup' clearance
    119–21
rain effect on electrical
    railtracks 187–8
sea foam 234
temperature effect on leaks
    52–3
tides 41–2
UK population support
    131–3
waves:
    capillary waves 173–5
    tidal effect on size 153
wax: recrystallisation 64
weight (mass):
    difference between equator
    and poles 142–3
    Earth's core 149–50
    flatulence effect 99–100
    moon effect 46
    MP3 players 209–10
whale strandings 243–5
wind:
    aircraft wing shape 184–6
    cloud effect on speed
    116–17
    noise on video 198–9
    sea foam 234
wood:
    branch characteristics
    128–30
    growth direction on planks
    127–8
    rotting vs burning 126–7
wool pilling 207–8
words: earliest known 100–1
World Wide Web: data volume
    218

X-rays: time telling on cloudy
    days 38

zebras: function of stripes
    3–4
zoopharmacognosy 18–19